Theory and Practice of Wood Pellet Production

T0171967

Zoltán Kocsis · Etele Csanády

Theory and Practice of Wood Pellet Production

 Springer

Zoltán Kocsis
Sopron, Hungary

Etele Csanády
Sopron, Hungary

ISBN 978-3-030-26181-8   ISBN 978-3-030-26179-5   (eBook)
https://doi.org/10.1007/978-3-030-26179-5

This Springer imprint is published by the registered company Springer Nature Switzerland AG
The registered company address is: Gewerbestrasse 11, 6330 Cham, Switzerland

Foreword

The production of pellets, as a renewable energy source, is continuously increasing worldwide. Industrial and agricultural wastes, by-products, and residues are available in all countries of the world, and one of the rational uses of these materials is the pellet making supply energy. Furthermore, a major advantage, using these by-products for energy production, contributes to the reduction of CO_2 emission to the environment.

This book is the first attempt to provide a comprehensive theoretical approach for the compaction process of various bulk materials prepared from wood wastes and by-products. A nonlinear rheological compaction model is developed and demonstrated with experimental results. A new similarity relationship is developed for the quick estimate of compaction energy as a function of the influencing variables. Furthermore, for the proper selection of channel length of the die ring a new approach is developed based on experimental measurements.

The practice of pellet making is equally treated in detail. There are many practical problems which fundamentally influence the production process. The preparation of the raw material, a continuous and accurate material flow to the press, the post-processing of pellets, and the economic analysis of the pellet production are important features needed to produce quality product at competitive price.

This new textbook on *Theory and Practice of Wood Pellet Production* may contribute to the better understanding of physical–mechanical processes associated with the compaction of bulk materials. The material is designed for undergraduate and postgraduate students, special courses at universities, research workers and application engineers working in the field of biomass utilization.

Sopron, Hungary György Sitkei
June 2019

Preface

Biomass as a renewable energy source is available in many countries worldwide, and it is important to utilize it economically and efficiently. The main sources are the wood industry, thinning and residues from forests, and agriculture with quite different material properties. Depending on their initial size or size distribution, moisture content, and plant origin, they require very different preprocessing and treatments to transform them into an appropriate final product. The most common practice is to produce pellets or briquettes and to utilize them for energy supply in households and power plants.

The whole production process for biomass conversion requires capital and energy investments, material and labor costs, transport and marketing activities which determine the production cost and the competitiveness of the product compared to other energy sources. In order to be competitive, a deep understanding of all processes associated with the whole production system is necessary. Especially, the utilization of agricultural by-products requires careful planning where a lot of problems are not fully solved yet.

This textbook is a first attempt to provide a systematic approach concerning the main processes of pellet production in theory and practice. The theoretical background of viscoelastic materials, their modeling, and calculation methods are discussed in detail. The general regularities of the compaction process, including a new nonlinear approach, are described and illustrated with measurement results. New measurement results for the energy consumption of pelletizing are provided, and also a generally valid similarity relationship is developed. Further new developments are the pressure—channel length relationship and the role of moisture in the development of bonding forces in the pellet.

The practice of pellet making has many technical and economic problems. The main working phases of pellet production including chipping, drying, post-chipping, and compaction are discussed. Further important questions are the energy balance of pellet production, the energetic utilization of pellets and the overall economy of pellet production.

The authors are indebted to Prof. György Sitkei for reading the manuscript and offering many useful suggestions, especially about the nonlinear treatment of the compaction process, to the development of similarity relations and the economic analysis of pellet production. The authors are also sincerely grateful to the staff of Springer Verlag for their excellent cooperation.

Sopron, Hungary   Zoltán Kocsis
                                                                                          Etele Csanády

Contents

Chapter 1
General Remarks

1.1 Introduction

The first introductory chapter covers the history of pellet making, first used for the stabilization of ingredients distribution in animal fodders. In our days the importance of pellet making is highly expanded to the field of biomass utilization for supplying energy.

The resources of biomass raw materials are abundant in all countries of the world which can economically be used either in households or industrial plants and for power generation. At the same time, biomasses may have quite different properties which require dissimilar preparation, compaction techniques and, mainly burning equipment.

1.2 History of Pellet Making

The history of pellets began in the 1870s. Both industrialization and animal husbandry at that time had undergone serious development. The growing feed requirements and the development of the industry, there was a need to produce feed containing a certain proportion of ingredients. The first machine was built in 1906, which was linked to the Louis Smulder & Co. machine factory in Utrecht, and offered a solution to the problem as compression kept mixing rates constant in the product. The pellet is actually a denser biomass.

In the 1920s, fuel pellets were first produced for fuel in North America. In the first period, sawdust from sawmills was used as raw material for making fuel pellets. The simple explanation for this is that sawdust can be directly converted into pellets without preparation. The compaction generated by the pressure is strengthened by fibrous cellulosic fibers. If the wood contains larger amounts of resins and oils, the product is less crumbly and less pressure is needed. Thus, with the use of simple technology, they were able to produce pellets from wood waste. The production of

© Springer Nature Switzerland AG 2019
Z. Kocsis and E. Csanády, *Theory and Practice of Wood Pellet Production*,
https://doi.org/10.1007/978-3-030-26179-5_1

more energy-intensive pellets using a wide range of raw materials was launched after 2000, typically in Western Europe and Canada. In 2012, 14.3 million tons of pellets were consumed annually in the EU Member States and this is expected to reach 50–80 million tons per year by 2020 (Marosvölgyi 2011).

The first systematic measurements on compaction processes of straw and hay have been made in the 1930s with suggestions for the basic pressure-volume density relationships (Pustigin 1937; Skalweit 1938). In the 1950s, the Poisson's ratio of straw and hay was investigated and stated that its value increases with higher densities. Furthermore, the resistance and counter-force of narrowing and curved pressing channels were also theoretically and experimentally investigated (Alferov 1956). Measurements in the high pressure range up to 1000 bar were performed in the early 1960s including the analysis of the briquetting and pelletizing processes (Osobov 1962; Dolgov and Vasilyev 1967; Schwanghart 1969). The use of rheological models for modeling of compaction processes was also published in this early time (Dolgov and Vasilyev 1967). A non-linear rheological approach was developed and tested later in 1990s (Sitkei 1994 and 1997). Combustion properties of biomass fuels including their pollutants were investigated in the last decades (Jenkins et al. 1998; Saastamoinen and Taipale 2000; Ryu et al. 2006). A detailed techno-economical analysis of wood pellets production was also given in our last decade (Pirraglia et al. 2010).

1.3   Importance of Pellet Making

Pellets are primarily competitive in the retail and municipal sector. They are perfect for serving small and medium-sized households. The pellet boiler is a technologically competitive alternative to gas boilers, as their automation is solved. Automation and programmability are an important aspect for the customer today. Another advantage is that storage of pellets requires fewer safety precautions, with lower risks than bottled gas and heating oil. The high degree of automation of transport, filling and dosing, especially in Western Europe, satisfies the customer's convenience. The space requirement can be one third smaller, than for the chopper, and this can be a serious argument for pellet burning in a family home. The amount of ash produced when using a good quality wood pellet that does not contain bark, does not exceed 2% so that emptying the ash is not too problematic and does not significantly reduce the comfort experience. Modern pellet boilers work at around 90–95% efficiency, which are more favorable than boilers with other fuels (Fenyvesi et al. 2008).

We have several aspects in the production of pellets. One important feature is the fixed geometric size. As a consequence the pellet becomes an easy-to-use fuel, and the combustion equipment can be programmed at a high level. Another aspect is the production of an appropriate density which is important in transportation and storage, but it is not negligible even in combustion technology. The required physical and combustion properties and the stability of the composition are important for the quality of the product.

Some advantages of pellet-based heat generation

- Heat regulation is solved,
- The pellet feeding can be automated,
- The pellet boiler has low maintenance,
- Wood pellets produce little ash (<1%),
- High energy efficiency (up to 95%),
- Eco-friendly,
- The energy density of pellets is high, the storage is convenient,
- Clean, renewable energy,
- Its heat output can be easily modified, varied,
- Low CO_2 emissions,
- Modern equipment at competitive prices,
- Well-defined, guaranteed heat value, well-scalable systems,
- Easy for ship.

1.4 Resources of Raw Materials

The possible sources of raw materials for pellet making are

- The forest from thinning, fallen branches and broken stems,
- Primary wood industry (saw mills) as sawdust, board edging residues,
- Secondary wood industry (furniture production) as chips, smaller and bigger pieces from sizing,
- Agriculture, mostly straws and stalks of various crops.

Pellet production is able to utilize widely available wood, forestry and agricultural waste and by-products. In principle, it is possible to make pellets from combustible, compressible solids and even partially utilize liquid fuels (e.g. fats, oils). It is important, however, to note that the properties of the base material seriously affect the product's quality, calorific value and on the boiler's life expectancy. The composition of the feedstocks has a serious effect on the amount of pollutants released, such as SO_x, NO_x, and the acids formed, as well as the melting of the resulting ash into slag, reducing the lifetime of the combustion device. One of the biggest sources of damage is the high chlorine content (Cl). The amount of chlorine in combustion influences the formation of dioxins. Dioxin causes serious health hazards, toxic to the human body. In addition to liver damage, it may also lead to renal cysteamination, which may result in characteristic neurodegenerative, immune, skin, gonadal, spleen and lymph node degeneration, bone marrow damage, and hormonal effects (such as uterine growth). The chlorine appears in the form of hydrochloric acid vapor in the flue gas. In addition to chlorine, nitrogen (N), potassium (K) and sulfur (S) content affect emission values. The maximum amount of nitrogen oxides (NO_x), sulfur oxides (SO_x) or acids produced is strictly regulated. These substances are highly detrimental to the environment, so it is not desirable to emit into the environment

(Fenyvesi et al. 2008). High concentrations of the flue gas must be cleaned from these materials, which is very expensive. If the raw material contains large amounts of silicon-containing powders, it will significantly increase the wear of the components of the production equipment. This is especially a problem when burning straw. The composition of the ash is important because if its melting point is not high enough, there is a risk that the ash will melt and deposition occurs. This will increase boiler corrosion, build up deposits, and larger quantities of blocked ash will clog the boiler and prevent its proper operation. Ash of agripellets has the lowest melting point. It is 700–900 °C, as opposed to the melting point of wood ash at about 1100–1300 °C. Therefore, agripellets should only be burned in a combustion plant designed for that purpose.

The moisture content is decisive for calorific value. The higher the moisture content is, the more energy is needed to heat the water and convert it to steam, thus the calorific value of the raw material will be significantly lowered. For higher efficiency, lower humidity is better, while the technology used in production requires some degree of moisture. High moisture content ($u > 13\%$) can result in clogging and pellet disintegration, lower moisture content below 10% requires higher pressure to obtain durable pellets. Based on the experience, therefore, the optimum moisture content for the production of pellets ranges from about 10 to 13% for wood pellets and 15%, for straw pellets. The temperature increase in pellet production increases the water loss and the moisture content of the pellet is therefore lower than that of the base material. In addition to the low moisture content, the other option is to use the condensation process to increase the energy recovery. In this process, condensation of the steam in the flue gas, recovers a part of previously introduced heat of evaporation. Gas boilers have a significant market share for condensing boilers but this is not yet the case for pellet boilers. It is true however, that pellet boilers operate with less air demand due to the much lower stoichiometric ratio of biomass (see Sect. 3.9). With condensation, the calorific value does not change, only the efficiency of the combustion plant.

Chapter 2
Theory of Compaction of Bulk Materials

2.1 Introduction

The mechanical properties of high pressure produced pellets depend on many factors. The most important factors are the type of wood, particle size, moisture content, the compaction pressure, the compaction speed, the deformation holding time, the punch diameter and the press temperature. Description of mechanical changes during high pressure compression of wood chips, the stress (σ)—deformation (ε) relationship are formulated by non-linear rheological methods, because the wood has a non-linear viscoelastic property. Consequently, during the compression process, the elastic modulus of wood chips greatly increases, and the resulted pellet at the end of the process suffers residual deformation. The amount of residual deformation determines the properties of a pellet, especially its density. This chapter gives a detailed overview of the rheological models and rheological equations, the non-linear compaction of bulk materials, some specific behavior of bulk materials with internal and external pores, general regularities of compaction and their energy requirements, pressure relations in the die-ring channel, similarity equation of the energy consumption of the pelleting process and the determination of channel length for a given pressure.

2.2 Wood as Rheological Material

Due to the typical structure of wood, the introductions of some definitions are needed for rheological discussion, which are not common in the mechanics of ordinary elastic bodies.

These definitions are the following

– *Biological yield point* This is the point on the stress-deformation curve at which the stress decreases or remains constant with increasing deformation. This point indicates the appearance of initial cell rupture in a small volume of a cellular system.

© Springer Nature Switzerland AG 2019
Z. Kocsis and E. Csanády, *Theory and Practice of Wood Pellet Production*,
https://doi.org/10.1007/978-3-030-26179-5_2

The yield point of biological materials plays an important part in determining their sensitivity to damage. If the load on a product does not reach the biological yield point, the cellular system will not be damaged, and spoiling of the product will not occur.

– *Rupture point* This is the point on the stress-deformation curve beyond which the stress decreases rapidly and significantly with increasing deformation. This point indicates failure over a significant volume of material. In soft, tough materials rupture occurs only after considerable plastic deformation.

– *Rigidity* The rigidity of a material is characterized by the tangent to the initial, more or less linear section of the stress-deformation curve. (This value is indeed nothing other than the modulus of elasticity). If the initial section of the curve is nonlinear, then either the initial tangent modulus, the secant modulus or the tangent modulus at a given point may be used (Fig. 2.1).

– *Degree of elasticity* This is the ratio of the elastic to the total deformation, when a material is loaded to a certain value and then unloaded (Fig. 2.1).

– *Toughness* The toughness is characterized by the work required to cause rupture in a material (mN/m^3), which is identical to the area under the stress-deformation curve.

– *Hardness* Hardness is characterized by the resistance of a material to penetration by an indenter.

– *Deformation work* The wood can store deformation energy in the elasticity range (Nm/m^3). If the deformation is elastic, the deformation work is provided by the area under the stress-deformation curve. If deformation is not elastic, it can be determined by adding the load-unloading cycle. Figure 2.1, the area under the CD unloading curve.

– *Mechanical hysteresis* Absorbed energy by the material in the load-unloading cycle. The energy gained is equal to the area between the curves. Mechanical hysteresis also characterizes the damping ability of wood.

Fig. 2.1 Explanation of basic definitions

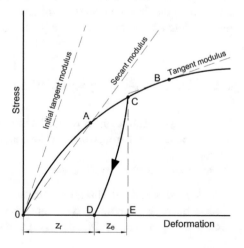

– *Energy Recovery* Relationship between energy recovered from unloading to the energy invested in load.

To clarify the rheology discussion, the substances should be divided into groups based on certain basic properties. Wide examination of the materials showed that elasticity, plasticity and viscosity are the three basic properties that characterize the rheological behavior of the materials. The three ideal bodies that exhibit these attributes are called Hooke's body (elasticity), Saint-Venant body (plasticity) and Newtonian fluid (viscosity).

Real materials are never perfectly elastic or plastic, so the three ideal bodies, as a standard, serve as a basis for comparison in our case when judging wood. The behavior of the ideal elastic body is shown in Fig. 2.2a. The stress is directly proportional to the elongation (symbol of the Hooke-body spring element), which is expressed by the well-known Hooke's Law. When the stress is eliminated, the elongation of the material ceases completely and the rebound occurs along the same line as the load. Such behavior is called "linear elastic".

Rubber recovers its original shape during unloading, but the curve is not linear (Fig. 2.2b). In this case we are talking about non-linear elasticity. The study of the compression of wood as non-linear viscoelastic material showed that these materials do not even have very small deformation at Hooke's elasticity properties (Mózes and Vámos 1968; Sitkei 1986). At the end of the unloading there is always a deformation of some value (residual deformation) (Fig. 2.2c). The degree of residual deformation determines the final density of the pellet.

The following basic relationships apply to elastic bodies. For tension or pressure, the elastic modulus (E) can be written according to Hooke's law as follows

$$E = \frac{\sigma}{\varepsilon} \tag{2.1}$$

where ε is the strain, expressed as $\varepsilon = \Delta l/l$, l—is the initial length and Δl—is the dimension change due to load (mm).

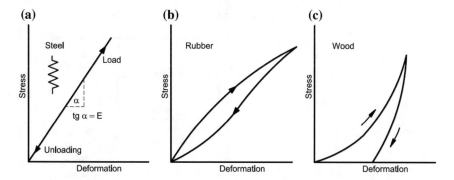

Fig. 2.2 Behavior of various bodies during the loading-unloading cycle

In case of shear stress, the given cross-section turns at an angle of θ and the shear modulus of elasticity (G) is

$$G = \frac{\tau}{\gamma} \qquad (2.2)$$

where

$\gamma = tg\theta$,
τ—shear stress.

Among the above moduli, taking into account the Poisson's ratio (v), the following relationship exists

$$E = 2G(1 + v) \qquad (2.3)$$

Most of the materials have a Poisson's ratio between 0.2 and 0.5. The value v = 0.5 is typical for liquids and rubber, meaning that the horizontal pressure in the enclosed space is equal to the vertical pressure. The other extreme case is when v = 0, such as cork.

Materials variously respond to a constant load, Fig. 2.3. Applying a constant stress for a time span t_1, elastic materials show a corresponding strain, which is also constant within the loading time. Plastic materials under high load show a small creep which corresponds to the compressibility of material. After removing the load, a plastic strain remains. In viscoelastic bodies, after an immediate elastic strain, a continuous creep occurs. After unloading the elastic strain is immediately recovered

Fig. 2.3 Different strain responses to a constant load

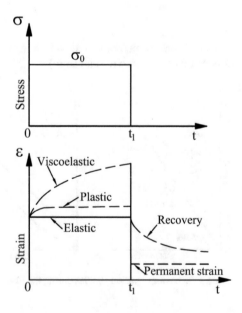

and a part of the creep strain recovers in time as a delayed recovery. The remaining part of the creep strain is the permanent strain.

The behavior of compressible bulk materials considerably varies as a function of load. With increasing load (pressure) the behavior shows increasing plasticity with increasing permanent strain. This basic behavior will be used in making pellets and briquettes.

The different behavior of materials is shown in a more detailed form in Fig. 2.4. Please note that a constant stress means a constant load only for elastic materials. With of viscoelastic materials, due to the creep phenomenon, an increasing load ensures constant stress. For similar reason, a decreasing stress corresponds to a constant strain as a function of time.

The behavior of wood is always different than the behavior of the ideal materials. Many studies have found that the stress (σ)—deformation (ε) correlation also depends on deformation velocity. This means that we do not have correlation between two variables (σ, ε) but between three variables (σ, ε, t). Materials for which time effects must be considered are called viscoelastic materials (Boltzmann 1876; Biot 1954). These materials have the properties of solid bodies (Hooke-spring) and partly of liquids (Newtonian liquids). For some materials and at relatively low loads, the relationship between stress and deformation is dependent on time only and not on the magnitude of stress. Such materials are referred to as linear viscoelastic materials. In wood, however, most of the deformation caused by the load cannot be recovered

Fig. 2.4 Behavior of materials **a** elastic, **b** creep, **c** relaxation, **d** combined creep and relaxation with permanent deformation

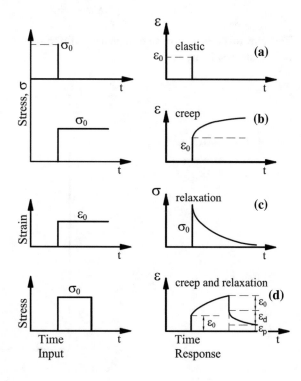

during the unloading process, so the relationship of stress-deformation depends on the magnitude of the stress over time. In this case we are talking about nonlinear viscoelasticity. The general theory of non-linear viscoelasticity is not yet elaborated, so we often have to make assumptions and apply the theory of linear viscoelasticity. The time-dependent behavior of the viscoelastic materials is described by the equations of the material law (constitutive equations), whose variables are stress (σ), deformation (ε) and time (t). Material laws for viscoelastic materials can be described on the basis of rheological models and by experimental data based on empirical correlations. The range of validity of rheological models should also be determined by experiments (Findley et al. 1976). The most commonly used experimental methods are creep and relaxation measurements.

Creep
Creeping means continuous deformation of the material under constant stress. Generally, three characteristic zones can be found in the creep process (Fig. 2.5). In the first phase of creep, the deformation velocity is degressive, the process is called primary creeping. In the second phase, the rate of deformation velocity is nearly constant, while in the third phase the rate of deformation velocity increases (progressive zone) and the process ends with rupture.

At any time t, the deformation (ε) is composed of the instantaneous elastic (ε_r) and creep deformation (ε_c)

$$\varepsilon_{(t)} = \varepsilon_r + \varepsilon_c \tag{2.4}$$

The rate of deformation is obtained by differentiation. Since $\varepsilon_r =$ constant, the deformation velocity is

$$\frac{d\varepsilon}{dt} = \frac{d\varepsilon_c}{dt} = \dot{\varepsilon} \tag{2.5}$$

Fig. 2.5 The stages of creep

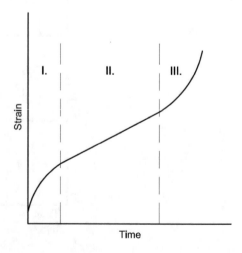

The time of each phase of the creep depends largely on the structure of the material and the stress. Therefore, a first step is to determine whether the first or the first and second stages of the creep curve are to be considered.

It should be noted that the creep behavior shown in Fig. 2.5 is valid first of all for tension and for non-confined compression. In the pellet making confined compression is always used and, in this case, the strain terminates at an asymptotic value.

Rebound

Loading a viscoelastic material with a constant stress, after an instantaneous elastic strain the deformation increases in time (creep). If the load is removed after a given time, the elastic strain is recovered instantaneously followed by recovery of a portion of creep strain at a decreasing rate (Fig. 2.6). Depending on material properties and on the applied stress, the recoverable strain and the remaining permanent deformation may be quite different.

The permanent deformation in pellet production plays a very important role ensuring the desired pellet density and durability. The moisture content is also an important influencing factor. Therefore, the process parameters (moisture content, pressure, loading velocity) should be selected so that the recoverable strain is as low as possible.

Relaxation

Another characteristic feature of viscoelastic materials is that, under constant deformation, the stress decreases gradually as a function of time (Fig. 2.7). The degree and velocity of the decrease depends on the structure of the material and the magnitude of deformation.

The decreasing stress generally holds asymptotically to a limit value. The rate of stress relaxation is characterized by relaxation time, which indicates the time during which the original stress value is reduced to $1/e$ (about 37%).

Using high pressure and compaction in pellet making, the portion of elastic strain decreases and, therefore, the stress relaxation will also be less. The behavior of the

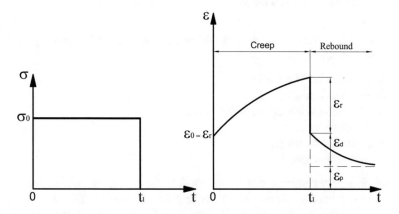

Fig. 2.6 The change of rebound as a function of time with permanent deformation

Fig. 2.7 The relaxation
progress in time

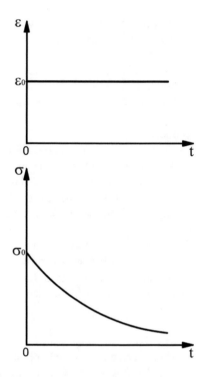

material approaches that of plastic bodies. As we will see later, the plasticity can also
be enhanced by temperature.

2.3 Modeling of Rheological Materials

2.3.1 Rheological Models

In the previous section we have seen that viscoelastic materials generally have the
following common phenomena

- instantaneous elastic behavior,
- creep under constant stress,
- stress relaxation under constant deformation,
- instantaneous recovery,
- time-dependent recovery,
- permanent deformation.

In the following we discuss mechanical models containing the basic elements of a
spring and a dashpot. The spring obeys Hooke's law with ideal elastic behavior and the
stress does not depend on the strain rate (Fig. 2.8). The dashpot as a damping element

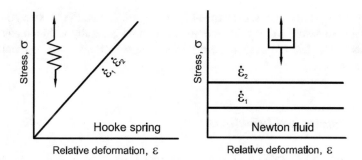

Fig. 2.8 Basic elements of rheological models

behaves differently from the spring the stress does not depend on the deformation but does depend on the deformation rate (Fig. 2.8).

Their behavior is given by the following simple equations

$$\sigma = E \cdot \varepsilon \quad and \quad \sigma = \eta \cdot \dot{\varepsilon} \tag{2.6}$$

where η is the coefficient of viscosity (Ns/m^2). The dashpot will be deformed at a constant rate when it is subjected to a step of constant stress

$$\varepsilon(t) = \frac{\sigma_0 \cdot t}{\eta} \tag{2.7}$$

In practice the spring and dashpot are combined in different fashions and numbers. The simplest two-element models are the Maxwell and Kelvin model, and their behavior is demonstrated in Fig. 2.9.

In the Kelvin model the spring and dashpot are connected in parallel, while in the Maxwell model they are connected in series. Neither of these simple models is suitable to describe the general behavior of viscoelastic materials. For example,

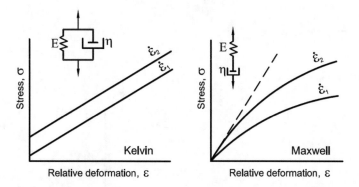

Fig. 2.9 Basic two-element rheological models and their response curves

the Kelvin model does not show a time-dependent relaxation, the Maxwell model is not suitable to describe creep. In order to obtain more generally valid models, more complicated models are needed. In the following we discuss several rheological models in detail.

2.3.2 Rheological Equations

The simplest model, which is most frequently used, is the three-element model, the two main types of which are shown in Fig. 2.10. The Maxwell and Kelvin models are respectively combined with a spring connected in parallel or serially. The model according to Fig. 2.10a may be described as

$$\sigma = E_1\varepsilon + T(E_1 + E_2)\frac{d\varepsilon}{dt} - T\left(\frac{d\sigma}{dt}\right) \tag{2.8}$$

where

$T = \frac{\eta}{E_2}$ is the relaxation time.

With a sudden load according to a step function, which is followed by constant deformation ($d\varepsilon/dt = 0$), the relaxation of the initial stress σ_0 may be calculated from the equation

$$\sigma(t) = \sigma_0 e^{-t/T} + \left(\frac{E_1}{E_1 + E_2}\right)\sigma_0\left(1 - e^{-t/T}\right) \tag{2.9}$$

For a sudden load and a subsequently constant stress ($d\sigma/dt = 0$), the variation of the deformation (i.e., the creep) is given by the equation

$$\varepsilon(t) = \left(\frac{\sigma_0}{E_1 + E_0}\right)e^{-t/T} + \frac{\sigma_0}{E_1}\left(1 - e^{-t/T}\right) \tag{2.10}$$

where $T = \frac{\eta}{E_\infty}$ and $E_\infty = \frac{E_1 \cdot E_2}{E_1 + E_2}$.

Fig. 2.10 The three-element models

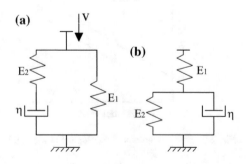

In practice a sudden load according to a step function cannot always be achieved so the calculation may be performed instead with loading at a finite constant rate v_0, or with loading corresponding to the motion of a crankshaft. When the deformation rate is constant v_0, the instantaneous strain is

$$\varepsilon = \frac{v_0}{L}t = at \quad and \quad \frac{v_0}{L} = \frac{d\varepsilon}{dt} = a \tag{2.11}$$

where

L—is the length of material.

With the latter relationship taken into account, the differential equation for the three-element model may be written as

$$\frac{d\sigma}{dt} + \frac{\sigma}{T} = \frac{E_1}{T}at + (E_1 + E_2)a$$

of which the solution is

$$\sigma(t) = E_1 at + TaE_2\left(1 - e^{-t/T}\right) \tag{2.12}$$

Equation (2.12) is valid in the time interval $0 < t < t_1$, where t_1 represents the end of the loading at a rate v_0. If the deformation remains constant after t_1, then the subsequent stress relaxation follows the equation

$$\sigma(t) = \sigma(t_1)e^{-t/T} + E_1\varepsilon(t_1) \cdot \left(1 - e^{-t/T}\right) \tag{2.13}$$

where $\sigma(t_1)$ and $\varepsilon(t_1)$ are the stress and strain at time t_1. After infinite time the stress tends to the value $\sigma_\infty = E_1\varepsilon(t_1)$.

The differential equation for the model shown in Fig. 2.10b is

$$\frac{d\sigma}{dt} + \frac{\sigma}{T} = E_1\frac{d\varepsilon}{dt} + \frac{1}{T}\frac{E_1 \cdot E_2}{(E_1 + E_2)}\varepsilon \tag{2.14}$$

where

$T = \eta/(E_1 + E_2)$.

With a sudden load and subsequent constant deformation, the stress relaxation may be calculated from the equation

$$\sigma(t) = \sigma_0 e^{-t/T} + \left(\frac{E_2}{E_1 + E_2}\right)\sigma_0\left(1 - e^{-t/T}\right) \tag{2.15}$$

For a sudden load and subsequent constant stress, the creep may be calculated from the equation

$$\varepsilon(t) = \frac{\sigma_0}{E_1} + \frac{\sigma_0}{E_2}\left(1 - e^{-t/T}\right) \tag{2.16}$$

where $T = \eta/E_2$.

With a constant deformation rate v_0 the differential equation assumes the form

$$\frac{d\sigma}{dt} + \frac{\sigma}{T} = E_1 a + \frac{1}{T}\left(\frac{E_1 \cdot E_2}{E_1 + E_2}\right)at$$

Introducing the asymptotic modulus $E_\infty = E_1 E_2/(E_1 + E_2)$, the stress-deformation equation will be

$$\sigma(t) = E_\infty at + Ta(E_1 - E_\infty)\left(1 - e^{-t/T}\right) \tag{2.17}$$

Equation (2.17) is valid in the time interval $0 < t < t_1$, where t_1 denotes the end of the loading period. The subsequent relaxation under constant deformation occurs according to the equation

$$\sigma(t) = \sigma(t_1)e^{-t/T} + E_\infty \varepsilon(t_1)(1 - e^{-t/T}) \tag{2.18}$$

The loading period t may be expressed in terms of the ratio of the deformation Δl and the loading rate v_0 and Eq. (2.17) may be expressed in the form

$$\sigma(t) = \varepsilon(t)\left[E_\infty + \frac{v_0 T}{\Delta l}(E_1 - E_\infty)\left(1 - e^{-\Delta l/v_0 T}\right)\right] \tag{2.18a}$$

In numerous cases the load is achieved by a crankshaft and the deformation varies with time according to the expression

$$\Delta l = r(1 - \cos \omega t) \tag{2.19}$$

while the loading rate is

$$v = r\omega \sin \omega t \tag{2.20}$$

The differential equation for the model shown in Fig. 2.10a may now be written in the form

$$\frac{d\sigma}{dt} + \frac{\sigma}{T} = \frac{E_1 r}{TL} - \frac{E_1 r}{TL}\cos \omega t + (E_1 + E_2)\frac{r\omega}{L}\sin \omega t$$

where

r—is the radius of the crank,
ω—is the angular velocity of the crank.

The solution of this differential equation is (Sitkei 1986)

$$\sigma(t) = \frac{\varepsilon_0}{2}\left[E_1 + E_2\frac{(T\omega)^2}{(T\omega)^2 + 1}e^{-t/T} + E_2\frac{(T\omega)^2}{(T\omega)^2 + 1}\sin\omega t - \left(E_1 + E_2\frac{(T\omega)^2}{(T\omega)^2 + 1}\cos\omega t\right)\right]$$

$$(2.21)$$

At the upper (dead) point, $\cos\omega t = -1$ and $\sin\omega t = 0$, with these values, the stress σ is

$$\sigma(t_1) = \frac{\varepsilon_0}{2}\left[2E_1 + E_2\frac{(T\omega)^2}{(T\omega)^2 + 1}\left(1 - e^{-t_1/T}\right)\right] \qquad (2.22)$$

where t_1 is the half-rotation time of the crank, or $t_1 = 180/6n$ and ε_0 is the strain corresponding to the stroke of the crank. Equations (2.21) and (2.22) are also valid for the model shown in Fig. 2.10b, if E_∞ is substituted for E_1 and $(E_1 - E_\infty)$ for E_2.

A frequently used model is the four-element Burgers model, shown in Fig. 2.11. The model consists of a Kelvin model connected in series to a spring and the dashpot (Mózes and Vámos 1968).

The model may thus be divided into three parts. The total deformation is given by the sum of the deformations of the individual parts, i.e.

$$\varepsilon = \varepsilon_A + \varepsilon_B + \varepsilon_C \qquad (2.23)$$

The stress is identical in all the parts, i.e.

$$\sigma = \sigma_A + \sigma_B + \sigma_C$$

The values of the individual stress components are

$$\sigma_A = E_0\varepsilon_A$$

Fig. 2.11 The four-element Burgers model

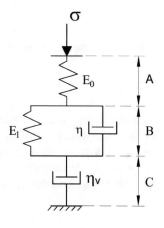

$$\sigma_B = E_r \varepsilon_B + \eta \dot{\varepsilon}_B$$
$$\sigma_C = \eta_v \dot{\varepsilon}_C \tag{2.24}$$

The above three equations yield the differential equation

$$\frac{d^2\varepsilon}{dt^2} + \frac{1}{T_r}\frac{d\varepsilon}{dt} = \frac{1}{E_0}\left[\frac{d^2\sigma}{dt^2} + \left(\frac{E_0}{E_r T_r} + \frac{E_0}{\eta} + \frac{1}{T_r}\right)\frac{d\sigma}{dt} + \frac{E_0}{T_r \eta_v}\sigma\right] \tag{2.25}$$

Equation (2.25) is suitable for describing both creep phenomena under a dead load and stress relaxation in linearly viscoelastic materials. For a constant load, Eq. (2.25) becomes simpler, as then $d\sigma/dt = 0$. The differential equation may be written in the form

$$\frac{d^2\varepsilon}{dt^2} + \frac{1}{T_r}\left(\frac{d\varepsilon}{dt}\right) = \frac{\sigma_0}{T_r \eta_v} \tag{2.26}$$

of which the solution is

$$\varepsilon(t) = \frac{\sigma_0}{E_0} + \frac{\sigma_0}{E_1}\left(1 - e^{-t/T_r}\right) + \frac{\sigma_0 t}{\eta_v} \tag{2.27}$$

where

$T_r = \frac{\eta}{E_1}$—is the retardation time.

The deformation rate is obtained by differentiating Eq. (2.27).

$$\frac{d\varepsilon}{dt} = \frac{\sigma_0}{\eta}e^{-t/T_r} + \frac{\sigma_0}{\eta_v}$$

Using this equation the deformation rate at $t = 0$ is

$$\dot{\varepsilon}(0) = \left(\frac{1}{\eta} + \frac{1}{\eta_v}\right)\sigma_0 = \tan\alpha$$

while at $t = \infty$ the deformation rate tends asymptotically to the value

$$\dot{\varepsilon}(\infty) = \frac{\sigma_0}{\eta_v} = \tan\beta$$

If the stress σ_0 at $t = t_1$ is terminated, the elastic portion of the deformation immediately disappears while the creep deformation decreases as a function of time and holds asymptotically to $\sigma_0 t_1/\eta_v$. The decay of deformation can be calculated using the superposition principle so that the tension $\sigma = -\sigma_0$ is superimposed at t moment $t = t_1$. This deformation value for $t > t_1$ times will be

$$\varepsilon(t) = \frac{\sigma_0}{\eta_v}t_1 + \frac{\sigma_0}{E_1}\left(e^{t_1/T_r} - 1\right)e^{-t/T_r} \tag{2.28}$$

Fig. 2.12 The four-element model behavior

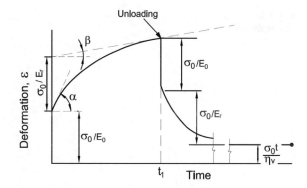

The behavior of the four-element model is shown in Fig. 2.12. From creep and relaxation tests the elements of the model can be determined and the angles of α and β can also be constructed (Fig. 2.12).

The Burgers model is now capable of describing residual deformation by means of a viscous element (η_v) bound to the model. At the same time, the problem with the viscous element is that it acts differently at different load speeds. So you always have to pick up different values depending on the speed. At low speeds, however, the model does not work because the viscous element does not gain strength. So this model can not be used to describe the compression processes of woody bulk materials.

The equations of the models discussed above have an exponential function. If the specific deformation ε is represented in a logarithmic coordinate system, then the exponential function can be represented by a single line. However, earlier research results showed that in logarithmic representation of wood we do not get a straight line for the relation $\varepsilon = f(t)$. This means that the modulus of elasticity is not constant but depends on the load (increasing its compression). In order to solve this problem, a generalized Kelvin model (Findley et al. 1976) was proposed. The generalized Kelvin model consists of Kelvin models and a spring and a viscous element connected in series. The first spring accounts for the momentary elastic deformation of the material, the n Kelvin model is characterized by delayed elasticity, whereas the viscous element finally corresponds to a permanent deformation (residual deformation) (Fig. 2.13). The equivalent of Eq. (2.27) for the generalized Kelvin model can be written in the following way

$$\varepsilon(t) = \sigma_0 \left[\frac{1}{E_0} + \frac{1}{E_{r1}} \left(1 - e^{-t/T_1}\right) + \frac{1}{E_{r2}} \left(1 - e^{t/T_2}\right) + \cdots + \frac{1}{E_{rn}} \left(1 - e^{-t/T_n}\right) + \frac{t}{\eta_v} \right] \quad (2.29)$$

where $T_1, T_2, \ldots, T_n$—the retardation or delay times of each Kelvin model
With the simplified writing, Eq. (2.29) can be written as follows

$$\varepsilon(t) = \sigma_0 \left[\frac{1}{E_0} + \sum_{i=1}^{n} \varphi_i \left(1 - e^{-t/T_i}\right) + \frac{t}{\eta_v} \right] \quad (2.30)$$

Fig. 2.13 The general
Kelvin model

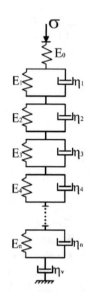

where $\varphi_i = \frac{1}{E_i}$—the reciprocal modulus.

If the number of Kelvin elements is infinite, the Σ sign can be written as an integral (apart from E_0 and η_v)

$$\varepsilon = \sigma_0 \int_0^\infty \varphi(T) \cdot (1 - e^{-t/T}) dT \tag{2.31}$$

where

$\varphi(T)$—is the distribution function of retardation times, or shortly the retardation spectrum.

The creep test of various materials has shown that experimental results can be easily approximated with the following equation

$$\varepsilon(t) = \varepsilon_0 + m't^n = \sigma_0 \left(\frac{1}{E_0} + mt^n \right) \tag{2.32}$$

It can be shown that this empirical function represents a specific retardation spectrum. Comparing Eqs. (2.31) and (2.32), shows that

$$mt^n = \int_0^\infty \varphi(T)(1 - e^{-t/T}) dT$$

which yields (Findley et al. 1976)

$$\varphi(T) = \frac{mnT^{n-1}}{\Gamma(1-n)}$$

The expression of the gamma function in the denominator is

$$\Gamma(1-n) = \int\limits_{0}^{\infty} \left(\frac{1}{T}\right)^{-n} \cdot e^{-t/T} d\left(\frac{t}{T}\right)$$

For most materials the exponent n is smaller than the unit, so in a logarithmic system of $\varphi(T) - T$ coordinates a negative inclination line system is obtained.

The generalized Kelvin-model is not suitable for describing the high pressure compression processes of wood because it is necessary to include relaxation times for each load. Calculating a larger number of elements considerably complicates the solution of the equation and thus affects the result obtained. In addition, the model still does not work at low speeds because of the viscous element in series, so it is not suitable for describing the remaining plastic deformation.

2.4 Non-linear Compaction with Plastic Deformation

The rheological model to be chosen should be able to describe all the possible processes occurring during a loading-unloading process. The loading may occur at any velocity between zero and infinity (different rate of straining or stressing), and therefore, the model must not contain a dashpot in series, since the dashpot does not offer any resistance if the loading velocity tends to zero. After the loading period relaxation or creep may be common and after an unloading the process is always associated with residual strain. The residual (permanent) strain will gradually be higher as the final value of the loading stress increases.

The simplest rheological model-, which may be chosen for compression processes is the three-element model, including friction elements describing plastic behavior (residual strain, Fig. 2.14. It is clear that only a non-linear approach makes it possible to vary the friction element from zero to a given value as the stress increases. The model is made up of a spring, a dashpot and friction elements all being a function of the instantaneous strain. The spring elements are given by

$$\sigma = E(\varepsilon) \cdot \varepsilon \quad and \quad \dot{\sigma} = E \cdot \dot{\varepsilon} + \dot{E} \cdot \varepsilon$$

The Kelvin-model, where the spring and dashpot are connected in parallel, is given by

$$\sigma = E(\varepsilon) \cdot \varepsilon + \eta \cdot \dot{\varepsilon} \quad and \quad \dot{\sigma} = E \cdot \dot{\varepsilon} + \dot{E} \cdot \varepsilon + \dot{\eta} \cdot \dot{\varepsilon} + \eta \cdot \ddot{\varepsilon}$$

allowing the variation of η as a function of load or strain.

Fig. 2.14 The 3-element
rheological model with
friction elements (Sitkei
1994)

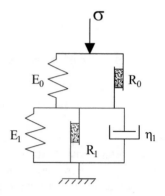

The friction element modeling plastic strain is also a function of strain with an initial value of zero and, after arriving at its maximum value, it remains constant independent of whether there is relaxation, creep or unloading. During the loading process, the stress-strain relationship is

$$\sigma = R(\varepsilon) \cdot \varepsilon \quad and \quad \dot{\sigma} = R \cdot \dot{\varepsilon} + \dot{R} \cdot \varepsilon$$

and the effect of the friction element is quite similar to those of the springs.
The differential equation of the above model can be derived in the following way. The two parts of the model are subjected to the same stress

$$\sigma = (E_0 + R_0) \cdot \varepsilon_1 \tag{2.33}$$

and

$$\sigma = (E_1 + R_1) \cdot \varepsilon_2 + \dot{\varepsilon}_2 \cdot \eta_1 \tag{2.33a}$$

Since the two parts are connected in series, the total strain is

$$\varepsilon = \varepsilon_1 + \varepsilon_2$$

and the strain rate becomes

$$\dot{\varepsilon} = \dot{\varepsilon}_1 + \dot{\varepsilon}_2$$

Taking the derivatives of Eqs. (2.33) and (2.33a) the following differential equation can be obtained

$$\ddot{\sigma} + \frac{\dot{\sigma}}{T_0} - \left[\frac{1}{T_0} \left(\frac{1}{T_1} + \frac{\dot{\eta}_1}{\eta_1} \right) + \frac{1}{T_1} \left(\frac{1}{T_1} + \frac{\dot{\eta}_1}{\eta_1} \right) - \frac{\dot{E}_1 + \dot{R}_1}{\eta_1} \right.$$

$$+\frac{\ddot{E}_0 + \ddot{R}_0}{E_0 + R_0} - \frac{2}{\eta_1}(\dot{E}_0 + \dot{R}_0)\left(1 + \frac{E_1 + R_1}{E_0 + R_0}\right)\Bigg]$$

$$= \left[\frac{2}{T_1}(\dot{E}_0 + \dot{R}_0) + \frac{1}{T_0}(\dot{E}_1 + \dot{R}_1) - \frac{1}{T_0}\frac{1}{T_1}(E_1 + R_1 + \dot{\eta}_1)\right] \cdot \varepsilon$$

$$+ 2(\dot{E}_0 + \dot{R}_0)\dot{\varepsilon} + (E_0 R_0)\ddot{\varepsilon} \tag{2.34}$$

where

$$T_0 = \eta_1/(E_0 + R_0) \quad and \quad T_1 = \eta_1/(E_1 + R_1)$$

If the coefficient of viscosity η_1 may be regarded as a constant, then a first order differential equation can be derived in the following form

$$\dot{\sigma} + \sigma\left(\frac{1}{T_0} + \frac{1}{T_1} - \frac{\dot{E}_0 + \dot{R}_0}{E_0 + R_0}\right) = \frac{E_0 + R_0}{T_1} \cdot \varepsilon + (E_0 + R_0) \cdot \dot{\varepsilon} \tag{2.34a}$$

For the linear case all the derivatives of the elements would become zero, and therefore, the above differential equations would have the forms

$$\ddot{\sigma} + \frac{\dot{\sigma}}{T_0} - \sigma\left(\frac{1}{T_0} + \frac{1}{T_1}\right) \cdot \frac{1}{T_1} = -\frac{E_0 + R_0}{T_1^2} \cdot \varepsilon + (E_0 + R_0) \cdot \ddot{\varepsilon} \tag{2.35}$$

or

$$\dot{\sigma} + \sigma\left(\frac{1}{T_0} + \frac{1}{T_1}\right) = \frac{E_0 + R_0}{T_1} \cdot \varepsilon + (E_0 + R_0) \cdot \dot{\varepsilon} \tag{2.35a}$$

Equations (2.34) and (2.34a) can only be solved numerically. However, Eqs. (2.35) and (2.35a) have the same analytical solution, if the model is subjected to a linearly increasing strain (an arbitrary loading velocity between zero and infinity) with

$$\varepsilon = a \cdot t \quad and \quad \dot{\varepsilon} = a \quad with \quad a = v/L$$

where

v—is the straining velocity,
L—is the sample height.

In this case Eqs. (2.35) and (2.35a) have the following solution

$$\sigma(t) = \varepsilon_\infty a t + a T (E_0 + R_0 - E_\infty) \cdot (1 - e^{-t/T}) \tag{2.36}$$

with

$$T = \frac{\eta_1}{E_0 + R_0 + E_1 + R_1} \quad and \quad E_\infty = \frac{(E_0 + R_0) \cdot (E_1 + R_1)}{E_0 + R_0 + E_1 + R_1}.$$

In order to use the above differential equations and their linear solution, the knowledge of the material law is needed in an appropriate form. Previous experimental investigations have shown that the confined compression of different agricultural materials may be approximated with Young's moduli of the general form (Sitkei 1981, 1994) (Fig. 2.23)

$$E_i = A_i \frac{\varepsilon^{n-1}}{(1 - \varepsilon)^n} \tag{2.37}$$

The friction elements and the coefficient of viscosity may also be approximated with equations similar to Eq. (2.37). These relationships may be determined experimentally including loading cycles with different loading velocities, and the subsequent relaxation and unloading cycle. The friction elements will be determined by the residual strains obtained at different straining levels and for several loading velocities.

Systematic numerical solution of Eqs. (2.34) and (2.34a) were performed and compared to those of the analytical solution according to Eq. (2.36) using the same model elements $E_i(\varepsilon)$, $R_i(\varepsilon)$ and $\eta_1(\varepsilon)$. These calculations have shown that the difference between the two methods does not exceed the ± 3 to 5%, including also the inaccuracy of the numerical procedure.

This finding may probably be explained by the fact that both methods give the same results as extreme values for the infinitely high and infinitely low loading velocities. This possible inaccuracy is fully acceptable for engineering calculations. Therefore, in the following we will use the more simple approximation method.

After the loading cycle, a constant load (σ_1) may be applied. As a result of creep, the deformation will increase as a function of time. The governing differential equation is given as Eq. (2.35a) taking $\sigma = \sigma_1$ and $\dot{\sigma} = 0$

$$\dot{\varepsilon} + \frac{E_1 + R_1}{\eta_1} \cdot \varepsilon = \left(\frac{E_0 + R_0}{\eta_1} + \frac{E_1 + R_1}{\eta_1} \right) \cdot \frac{\sigma_1}{E_0 + R_0}$$

With the initial conditions $\sigma = \sigma_1$ and $\varepsilon = \varepsilon_1$ at $t = t_1$, the following solution may be obtained

$$\varepsilon(t) = \varepsilon_1 \cdot e^{-(t-t_1)/T_1} + \sigma_1 \left(\frac{1}{E_0 + R_0} + \frac{1}{E_1 + R_1} \right) \cdot \left(1 - e^{-(t-t_1)/T_1} \right) \tag{2.38}$$

Since E_0, E_1 and R_0, R_1 are the function of strain, Eq. (2.38) can be solved by iteration.

If the loading stress is removed at time t_2 and at strain ε_0, first an instantaneous elastic recovery occurs followed by creep recovery at a decreasing rate. Keeping in mind that R_0 remains constant, the instantaneous elastic recovery is due to the pressure difference corresponding to $(E_0 - R_0)$.

$$-\Delta\varepsilon = \varepsilon - \varepsilon_0 \quad if \quad E_0(\varepsilon) = R_0(\varepsilon)$$

where E_0 means the final strain before unloading. If the compression curve would be linear, the elastic recovery would be

$$-\Delta\varepsilon_e = \frac{E_0 - R_0}{E_0}\varepsilon_1 \tag{2.39}$$

where ε_1 is the strain of the spring E_0. Sadly, it is not the case and, therefore, correction should be made to take the varying tangent of the compression curve into account, Eq. (2.48) describes the pressure-strain relation and its derivation with respect to strain gives the tangent of the compression curve. Processing of experimental results with beech chips gave the following correction

$$-\Delta\varepsilon_e = \frac{(E_0 - R_0)/E_0}{B\left(\frac{\varepsilon}{1-\varepsilon}\right)^{n-1}/(1-\varepsilon)^2}\varepsilon_1 \tag{2.39a}$$

with the constants $B = 0.11$ and $n = 1.15$.
In similar manner, the creep recovery of the extended Kelvin-model is

$$-\Delta\varepsilon_c(t) = \frac{E_1 - R_1}{E_1} \cdot \varepsilon_2(1 - e^{-(t-t_2)/T_1}) \tag{2.40}$$

with $T_1 = \frac{\eta}{E_1+R_1}$ and the resulting strain will be expressed by the relation

$$\varepsilon(t) = \varepsilon_0 - \Delta\varepsilon_e - \Delta\varepsilon_c(t) \tag{2.41}$$

where ε_2 means the strain component of the Kelvin-part.
To use Eqs. (2.39), (2.40) and (2.41) the strain components ε_1 and ε_2 must be known. Equating Eqs. (2.33) and (2.33a), the following differential equation is obtained

$$\dot{\varepsilon}_2 + \frac{\varepsilon_2}{T} = \frac{\varepsilon}{T_0}$$

and using the linear straining with $\varepsilon = a \cdot t$, the ε_2 strain component has the following time-dependent value

$$\varepsilon_2(t) = \frac{T}{T_0}a \cdot t - \frac{a \cdot T^2}{T_0}(1 - e^{-t/T}) \tag{2.42}$$

and $\varepsilon_1 = \varepsilon - \varepsilon_2$.
Equation (2.40) has the same problem as Eq. (2.39) but the rebound of the Kelvin part is generally small compared to the elastic rebound, Table 2.1.

    In many cases a single relaxation time is not enough to describe time-dependent processes with acceptable accuracy, especially for long times. In this case the model should include several Kelvin-parts in series which can be handled in the same way as shown before. A simpler method is to use the spectrum of relaxation times with

Table 2.1 Elastic and delayed rebound of beech wood

Strain	0.5	0.65	0.8
Pressure, bar	115	325	1000
Elastic rebound, %	22	10.8	3.25
Kelvin rebound, %	4	1.6	0.4

the following empirical equation

$$T = T_k + a \cdot (t - t_1)^n \tag{2.43}$$

where

T_k—is the minimum relaxation time at the beginning of the loading cycle,
a—is a constant,
t—is the beginning of the loading time,
t_1—is the end of loading time,
n—is an exponent.

As an example, Table 2.2 shows the constants in Eq. (2.37) using beech sawdust in the pressure range up to 1000 bar.

The compaction behavior of different materials is very similar to each other. Figure 2.15 shows compression curves as a function of relative deformation for infinitely high and low loading speeds. Relative values of creep and relaxation are also given, in which $\Delta\sigma = \sigma_0 - \sigma_\infty$ for a given ε value and $\Delta\varepsilon = \varepsilon_\infty - \varepsilon_0$ for a given stress value.

When pressure increases, the relative values of creep and relaxation are gradually reduced. Since plastic deformation is an important parameter for the compaction process, we have investigated it both theoretically and experimentally. After a given pressure, the load is removed and part of the deformation is recovered (rebound) which decreases the density corresponding to the applied pressure. Figure 2.16 shows calculated and experimentally measured relative rebounds for beech sawdust as a function of maximum strain. The experimental results and their scattering agree well with the theoretically calculated ones. The results support the effect of loading velocity on the rebound, especially for decreasing end strain.

By compaction of particleboard, different particle sizes of different species of wood are used. A closer examination of the pressure curves produced the surprising result that for a given compactness (in this case 710 kg/m³), the harder black locust (*Robinia pseudoacacia*) needs less pressure compared to spruce. The experimentally recorded pressure curves are shown in Fig. 2.17 in a logarithmic coordinate system.

Table 2.2 Constants in Eq. (2.37)

	E_0	R_0	E_1	R_1	η_1
A_i	95	64	70	60	1580
n	1.4	1.55	2.3	2.4	1.4

Fig. 2.15 Beech sawdust compression curves in the briquetting/pelletizing range

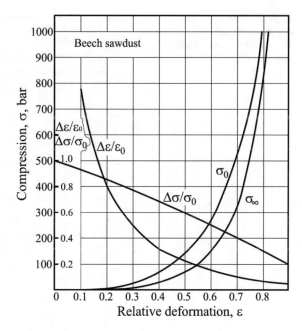

Fig. 2.16 Relative rebound of beech sawdust depending on maximum deformation

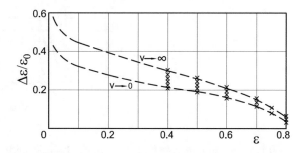

The physical and mechanical properties of the particles are used to explain this phenomenon. When pressing granular particles, the reduction of volume occurs due to the compaction in the outer pores (gaps) between the particles and the pores inside the particles. In the first stage of compression, the outer pores are consumed, and gradually the volume of the inner pores is also reduced. In accordance with the internal pore distribution, the initial bulk density of each species also varies considerably. The solid volume density of black locust is 660 kg/m^3 and therefore to achieve the final density (710 kg/m^3) it should be compressed to a lesser extent. Spruce has a starting volume density of 340 kg/m^3, so eliminating the outer pores is far from sufficient to reach the desired final density. When both materials will be further compressed (900 kg/m^3), the harder black locust needs the greater pressure. The decrease of the inner pores, depending on the volume density, is shown in Fig. 2.18 for the two wood species.

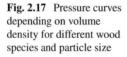

Fig. 2.17 Pressure curves depending on volume density for different wood species and particle size

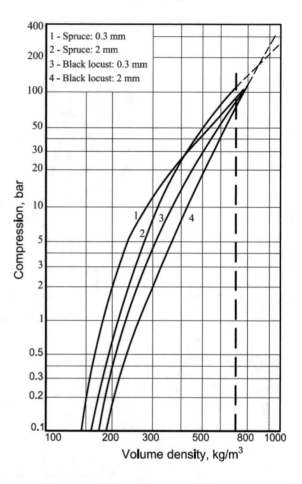

It is interesting to note that if the initial bulk density is compressed to the initial volume density of the particles, 40–43% of the internal pores are removed for both species. That means that the outer pores do not disappear completely, while the inner pores close to the boundary surface can easily be removed. This explains the relatively large reduction of inner pores.

The process can be depicted in a triangle diagram (Fig. 2.19). Every point within the triangle clearly defines the relationship between compactness, inner and outer pores. The course of the curves is characteristic for the compression course. The top tip of the graph corresponds to the absolute solid wood, so the compactness is a relative value. The density of the absolute solid wood is approx. 1500 kg/m^3. If only the outer pore would change during compression, the volume density of the original wood would be reached (dotted lines from "1" and "2" to the right) (Sitkei 1994).

The non-linear compaction model discussed above is suitable to determine the effect of deformation speed on the pressure development. Figure 2.20 shows these calculations for compaction of beech sawdust (Sitkei 1994). A loading speed of 100 cm/s

Fig. 2.18 Changes in the internal pores as a function of bulk density during compression

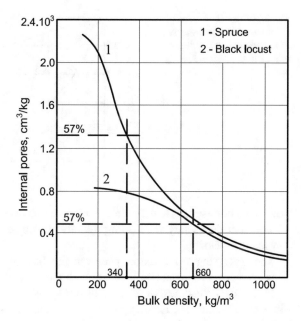

Fig. 2.19 The triangle diagram. Interrelation between density and internal, and external pores 1—Spruce, bulk density 140 kg/m^3, solid wood 340 kg/m^3, internal pores 77.33%, external pores 13.34%, 2—Black locust, bulk density 180 kg/m^3, solid wood 660 kg/m^3, internal pores 56%, external pores 32%

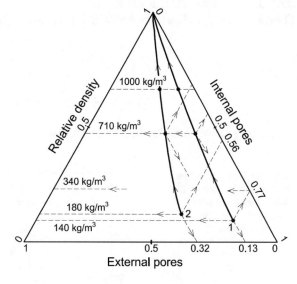

Fig. 2.20 Effect of speed of deformation on the pressure development between the infinite high and low loading cases

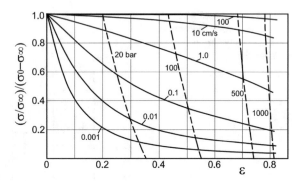

practically corresponds to the infinitely high velocity without any viscoelastic effect. On the other side, a very low loading velocity of 0.001 cm/s approaches the infinitely low loading velocity.

The processing of relaxation measurements have shown that the relaxation time does not remain constant but increases with the elapsed time.

Figure 2.21 illustrates measurement results using spruce chips with an average diameter of 2 mm. The loading pressure, which is followed by relaxation, considerably influences the relaxation time.

The curves may well be described with Eq. (2.43) as follow

$$T_1 = 4 + 1.1 \cdot (t - t_1)^{0.75}$$

and

$$T_2 = 1.7 + 0.75 \cdot (t - t_1)^{0.7}$$

which may conveniently be used in calculations.

Fig. 2.21 Relaxation time distribution for spruce chip with geometric mean of 2 mm for two different loading pressure

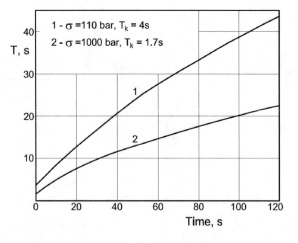

2.5 General Regularities of Compaction Processes

The time course of the stress and compression processes is shown in Fig. 2.22. After loading, the wood will usually remain on a constant load (σ_0) for a shorter period of time ($t_2 - t_1$) and creep occurs during this time (ε_1). In this case the creeping means further compaction of the material under constant stress (pelletizing pressure) (Sitkei 1981, 1994).

As the load increases, the modulus of elasticity also increases, i.e., the process is characterized by the initial (E_0), instantaneous (E) and the secant modulus (E_k). As a result of the compression process, the material will have a smaller volume and its elasticity modulus will increase considerably. Therefore, the laws of linear viscoelasticity can not be used to describe compression processes.

Investigations on the compaction of straw and hay were already carried out in the 1930s and exponential and parabolic functions were suggested to describe the pressure—volume density relationship (Pustigin 1937; Skalweit 1938), (Fig. 2.23).

The exponential relationship is used in the form

$$p = K\left[e^{a(\rho - \rho_0)} - 1\right] \tag{2.44}$$

where K and a are material-dependent constants. For forage materials with 10–20% moisture content, $K = 5.2 \times 10^{-3}$ and $a = 5 \times 10^{-2}$ if the initial volumetric density ρ_0 is less than 80 kg/m^3 (Schwanghart 1969)

Parabolic relationships are generally used in the forms

$$p = C(\rho - \rho_0)^m \tag{2.45}$$

where

ρ_0—is the initial volume density,
C and m are material-dependent constants.

Fig. 2.22 Creep and rebound of the compressed material

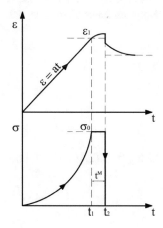

Fig. 2.23 Compaction of
material under a piston in a
cylinder

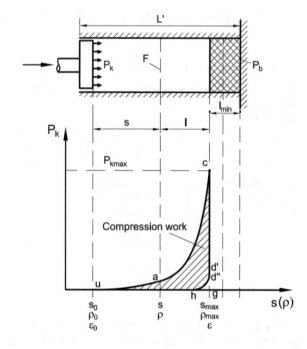

The term $(\rho - \rho_0)$ can be replaced by the strain ε which is more convenient to use in the processing of experimental results (Sitkei 1981). Using a press cylinder, the volumetric density varies according to the following equation.

$$\rho = \frac{\rho_0}{1-\varepsilon} \quad and \quad (\rho - \rho_0) = \rho_0 \frac{\varepsilon}{1-\varepsilon} \tag{2.46}$$

In the pellet production the relative rebound of the pellet after unloading is an important parameter determining the final density of the pellet. It may be expressed either with the relative strain decrease $\Delta\varepsilon/\varepsilon$ or with the relative density decrease $\Delta\rho/\rho$. Between these quantities, using Eq. (2.46), the following interrelation holds

$$\Delta\varepsilon = \frac{\rho_0}{\rho^2}\Delta\rho$$

or

$$\frac{\Delta\varepsilon}{\varepsilon} = \frac{\rho_0}{\rho-\rho_0} \cdot \frac{\Delta\rho}{\rho} \tag{2.46a}$$

Using Eqs. (2.46), (2.44) and (2.45) can be rewritten in the form

$$\sigma = K\left\{e^{A\rho_0\frac{\varepsilon}{1-\varepsilon}} - 1\right\} \tag{2.47}$$

$$\sigma = A\left(\frac{\varepsilon}{1-\varepsilon}\right)^n \tag{2.48}$$

The instantaneous elastic modulus can be calculated from the following equation

$$E = \frac{\partial\sigma}{\partial\varepsilon} = \frac{K \cdot A \cdot \rho_0}{(1-\varepsilon)^2} \cdot e^{A\rho_0\frac{\varepsilon}{1-\varepsilon}} \tag{2.49}$$

and the initial value of the elastic modulus is

$$E_0 = K \cdot A \cdot \rho_0 \tag{2.50}$$

The average value of the elastic modulus can also be defined (secant modulus) as follows

$$E_k = K\left\{e^{A\rho_0\frac{\varepsilon}{1-\varepsilon}} - 1\right\}/\varepsilon \tag{2.51}$$

The instantaneous modulus of elasticity in Eq. (2.48) is given by

$$E = \frac{\partial\sigma}{\partial\varepsilon} = A \cdot n\left(\frac{\varepsilon}{1-\varepsilon}\right)^{n-1}/(1-\varepsilon)^2 \tag{2.52}$$

or the secant modulus

$$E_K = A\frac{\varepsilon^{n-1}}{(1-\varepsilon)^n} \tag{2.53}$$

The exponent n depends on the size and strength of particles and on the pressure range. Its value varies between 1.5 and 2.5 for various wood chips (Sitkei 1994). Especially the strength of chips influences the course of the compaction curve as a function of displacement. Soft particles can be compacted at relatively low pressures and, increasing further the pressure, the curve ascends steeply giving a higher exponent. On the contrary, particles of hardwood species with higher individual strength will be compacted more uniformly as a function of displacement resulting in lower exponent n. Due to the anatomical structure of wood, the individual strength of particles depends not only on wood species but also on the size of particles. With decreasing particle size the surface—volume ratio increases (Csanády and Magoss 2013; Akdeniz and Haghighat 2013).

The surface of a particle is created by fracture with a highly uneven surface. The load-bearing capacity of this near-surface layer is less compared to that of the sound inner part. As a consequence, a smaller particle may behave similarly as it would be originated from a softer wood species. The size effect on the compaction process can only be determined experimentally.

The exponential relationship, Eq. (2.44) or (2.47), can be used also in the following modified form. We can define the density ratio (DR) as

$$DR = \frac{\rho - \rho_0}{\rho_m - \rho_0}$$

where ρ_m means the true density of wood, $\rho_m \cong 1500$ kg/m^3.
The density ratio is determined by the applied pressure according to the following equation

$$DR = 1 - e^{-(p/p_0)^n}$$

where the characteristic pressure p_0 corresponds to a pressure value at which the density ratio is $(1 - 1/e) = 0.6321$. Using Fig. 2.15 for a loading velocity of 1 cm/s, $p_0 = 800$ bar and $n = 1.2$ are obtained.

2.5.1 Energy Requirements of Compaction

The energy requirements of compaction by balers or pelleting machines comprise the net compaction work and the work spent in pushing. The net compaction work may be determined most simply by means of a pressing cylinder (Fig. 2.24). A piston moving from position s_0 compresses the material so that its volumetric weight ρ_0 increases

Fig. 2.24 Compression and pushing work at pressing

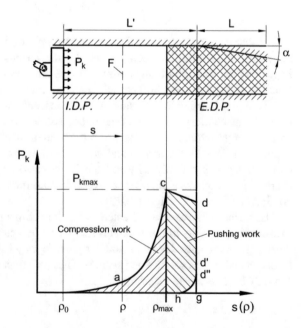

to the value ρ_{max}, while the pressure acting on the piston gradually increases. The work spent in compaction is given by the area under the pressure curve

$$A_c = F \int p \, ds \qquad (2.54)$$

or, related to the unit mass

$$\frac{A_c}{M} = \frac{F}{M} \int p \, ds \qquad (2.55)$$

where

$M = FL\rho_0$,
F—is the area (cross-section),
L—is the material height before pressing.

Alternatively, since $M = FL\rho_0$, the using of relations

$$\frac{\rho_0}{\rho} = \frac{L}{L - s}$$

or

$$-ds = L\rho_0 d\left(\frac{1}{\rho}\right)$$

yields the equation

$$\frac{A_c}{M} = \int_{\rho_0}^{\rho} p \, d\left(\frac{1}{\rho}\right) \qquad (2.56)$$

i.e., if the compression curve is plotted using a system of $p - 1/\rho$ coordinates, the area under the curve gives the specific net compression work.

In continuous pressing processes the material is compressed not in a closed space but in a suitably designed channel. Counter support of the material is ensured by narrowing of the channel and by wall friction. In this case the net compression work is accompanied by pushing work, which must be added to it, as shown in Fig. 2.24.

The pushing work (A_s) may be calculated on the basis of Eq. (2.56), with the assumptions that the pressure $p = p_{max} = \text{const}$ (disregarding the slight variation over the section c–d), and that $1/\rho = 1/\rho_{max} = \text{const}$. For these values

$$\frac{A_s}{M} = \frac{p_{max}}{\rho_{max}} \qquad (2.57)$$

and the total specific work is

$$\frac{A}{M} = \frac{A_c}{M} + \frac{A_s}{M} = \int_{\rho_0}^{\rho} p d\left(\frac{1}{\rho}\right) + \frac{p_{max}}{p_{max}} \tag{2.58}$$

The specific work spent in compression depends primarily on the pressure range and the moisture content. Figure 2.25 shows the specific compression work as a function of moisture content for various final pressures, for meadow grass and alfalfa (Sacht 1967).

The energy requirements are also affected by the velocity of compression. For increasing velocity the pressure required to attain a given volumetric weight generally increases, and together with it the energy requirements. The ratio of the compression and pushing work also depends crucially on the dimensions of the compression channel. The compression work is independent of the channel dimensions, but the pushing or friction work decreases significantly as the cross-sectional area of the channel increases.

The performance requirements of the pelletizing machines are similar. Making small-diameter pellets is very energy-intensive due to the increase in friction work. Figure 2.26 shows the power requirements for pelleting fodder flour, as a function of pellet diameter (Schwanghart 1969).

The curve may be described by the empirical equation

Fig. 2.25 Specific compaction work as a function of moisture content for meadow grass and alfalfa

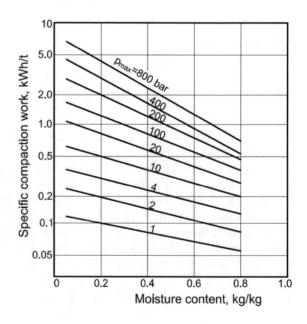

Fig. 2.26 Specific energy requirements of pelleting process as a function of pellet diameter for fodder flour

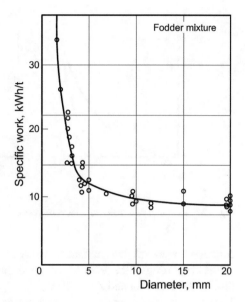

$$\frac{N}{Q} = \frac{d}{0.09(d - 1.5)} \qquad (2.59)$$

where

Q—is the throughput of the machine, t/h,
D—is the diameter of the pellets, mm.

2.5.2   Rebound of Material After Compaction

Materials are compressed by a pressure p to a volumetric density ρ_{max} during compression. After unloading the material rebounds according to its viscoelastic properties, i.e., its volumetric weight decreases to a value ρ_k (Fig. 2.27a). The equation for the unloading curve may be given, similarly to Eq. (2.44), as

$$p = K_1\left(e^{a_1(\rho - \rho_k)} - 1\right) \qquad (2.60)$$

where K_1 and a_1 are constants depending on the type of material, its moisture content and the period of time spent under the maximum pressure.

Unloading may be performed either immediately after the value ρ_{max} is attained, or after a certain time t^* has elapsed (Fig. 2.27b). The longer the time t^*, the less the extent of rebound of the material, and the higher the value of ρ_k.

The volumetric weight ρ_k corresponding to the stable state is not attained immediately, even in the case of instantaneous unloading. Rebound of a material may

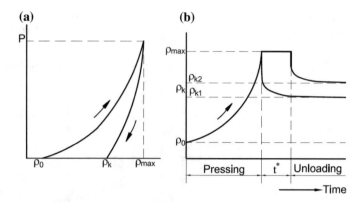

Fig. 2.27 Rebound of a material after compression

continue for several days, naturally with very small increments. However, a considerable proportion of the rebound usually occurs within a short time after unloading, Fig. 2.28 (Mohsenin and Zaske 1975).

The decrease in volumetric weight during rebound may be calculated from the empirical relationship

$$\rho_k = a \cdot \rho_0 + b \cdot \rho_{max} + ct^*(\rho_{max} - \rho_0) \tag{2.61}$$

where the constants may be substituted by the values $a = 0.4$, $b = 0.57$ and $c = 4.6 \times 10^{-3}$ for forage materials, for the pressure range encountered in balers. The value t^* must then be substituted in s units, and the range of validity for the constant c is $0 \leq t^* \leq 60$ s.

Fig. 2.28 Rebound of alfalfa wafers as a function of time

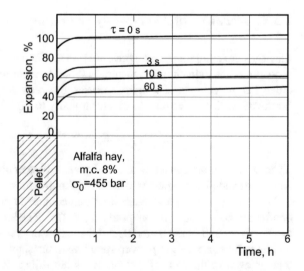

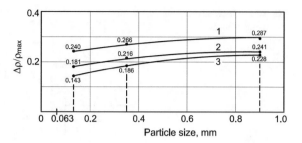

Fig. 2.29 Relative rebound of different wood species as a function of particle size. p = 1400 bar. 1—spruce, 2—oak, 3—black locust

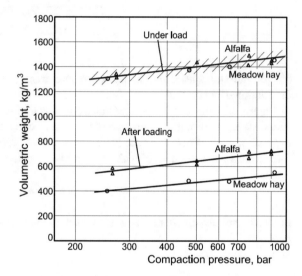

Fig. 2.30 Rebound of alfalfa pellets as a function of pressure

Figure 2.29 shows the relative rebound of different wood species as a function of particle size pressing pellets of 6 mm diameter at a pressure of 1400 bar and ambient temperature. Hardwoods have smaller rebound while bigger particle size slightly increases the rebound resulting in smaller final pellet density.

Figure 2.30 illustrates the rebound of alfalfa and meadow grass with 17–20% moisture content for the pressure range encountered in pelleting machines, as a function of maximum pressure, after three days' storage (Matthies and Busse 1966).

2.6 Pressure Distribution in a Compaction Chamber

During the compression of materials both in cylinders and channels, the pressure in the space before the piston is not uniform. Accordingly, the density (volumetric weight) of the material varies as a function of the distance from the piston. The reason for this is that the material is exposed to friction along the walls surrounding

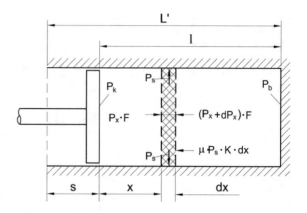

the compression space, and the axial pressure arising in the material is reduced by
the frictional force.

Figure 2.31 illustrates the force relations developing in a compression cylinder.
At a distance x from the piston a pressure p_x acts, creating a lateral wall pressure p_s
depending on the Poisson's ratio (v). Knowing the friction coefficient, the frictional
force acting on an element of width dx may be calculated as

$$S = \mu \cdot p_s \cdot D \cdot \pi \cdot dx \tag{2.62}$$

In the case of unidirectional loading, the relationship between the pressures p_x and
p_s is

$$p_s = \frac{v}{1-v} p_x \tag{2.63}$$

and so equilibrium of the element dx may be expressed by the differential equation.

$$dp_x F + \mu \frac{v}{1-v} p_x D\pi\, dx = 0 \tag{2.64}$$

The solution of this equation is

$$p_x = p_k e^{-kx} \tag{2.65}$$

where

$$k = \frac{4}{D} \mu \frac{v}{1-v} \tag{2.66}$$

The pressure on the base of pressing cylinder is obtained by substituting the length
l for x

Fig. 2.32 Pressure distribution before a pressing piston

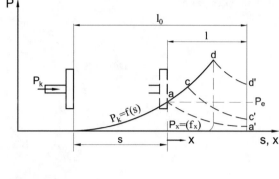

$$p_b = p_k e^{-kl} \tag{2.67}$$

Figure 2.32 shows schematically the pressure curves for various degrees of compression. In order to apply Eq. (2.65), the Poisson's ratio and friction coefficient μ of the material being compressed must be known as functions of the pressure and moisture content. The Poisson's ratio of forage materials increases with their volumetric weight and with their moisture content. For the pressure range encountered in balers $v = 0.25 - 0.35$, while for pelleting machines, $v = 0.35 - 0.45$ (Alferov 1956; Sacht 1967).

The coefficient of friction is a function of the moisture content but the pressure also has a significant effect on it. This latter effect is manifested by the fact that large pressure deforms the wood so that the contact surface is modified and that water is compressed from the material which reduces the friction coefficient. As a result, at high pressures the friction coefficient decreases in the higher moisture range.

2.7 Pressure Relation of the Pellet Production

The agricultural materials are pelleted mostly in die rings, whose layout is illustrated in Fig. 2.33. The farinaceous material fed is compressed by a suitably adjusted roller (similar to that of a chaser mill) and pressed into boreholes of the ring.

The compressed material emerging from the boreholes is cut by a knife on the outer side. Figure 2.34 shows the deformation conditions for the material under the roller (Schwanghart 1969). The roller is set relative to the die ring with a gap y_r, and so a compressed layer (carpet) is formed on the running surface of the ring, whose thickness y_t exceeds the value y_r, due to rebound of the material. The layer thickness s of fresh material fed before the roller is gradually reduced as a result of compression by the latter. With increasing pressure the initial thickness, s' of the layer decreases gradually until it reaches the minimum value y_r. At point b the pressure attains its maximum value and the material is pressed by the roller into the extrusion channels. The pressure remains practically constant in the material during pushing and then decreases rapidly as the gap increases.

Fig. 2.33 Principles of
pellet production

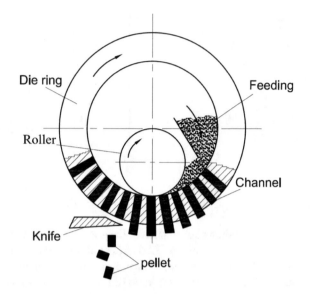

Fig. 2.34 Deformation of
material under the roller of a
pelleting machine

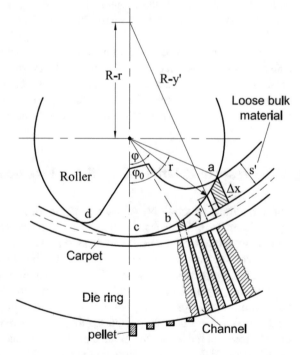

Fig. 2.35 Pressure distribution between the roller and die ring in pressing fodder flour

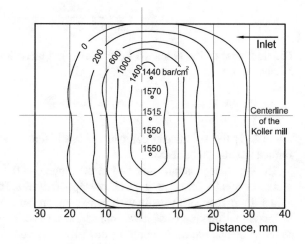

Figure 2.35 shows the distribution of pressure between the roller and die ring during the compression of fodder flour (Schwanghart 1969).

The width of the roller is 50 mm, the inner diameter of the ring is 300 mm and the diameter of the boreholes is 4.5 mm. The compressibility of floury materials may be described by an equation of the form

$$p = e^{A\rho + B} - D \tag{2.68}$$

while for straw-grit-molass mixtures the following equation is used

$$p = a\rho^b \cdot e^{c\rho} \tag{2.69}$$

The constants appearing in the equations are determined experimentally. With regard to the fact that the volumetric weight increases proportionally to the decrease in layer thickness, i.e.

$$\frac{\rho}{\rho_0} = \frac{y_0}{y} \tag{2.70}$$

it is possible to write

$$\frac{y}{y_0} = 1 - \varepsilon = \frac{A\rho_0}{\ln(p + D) - B} \tag{2.71}$$

where D and B are materials depending constants.

For example, for the pelleting of fodder flour ($\rho_0 = 500$ kg/m^3), the following relationship was found experimentally (Schwanghart 1969)

$$\frac{y}{y_0} = \frac{5.05}{\ln(p + 0.453) + 5.842}$$

The thickness of the unloaded layer remaining after the roller has passed may be calculated as

$$y_t = y_r + \frac{p_{max}}{C}$$

where C is the volumetric compaction coefficient, whose value for fodder flour amounts to 30,000 bar/cm.

The gap y_r is generally selected in the range 0.4–0.8 mm, since the throughput decreases rapidly for wider gaps (Schwanghart 1969). The modulus of elasticity of fodder flours increases steeply with increasing pressure, as shown in Fig. 2.36.

Individual amounts of relative compaction have been marked on the curve. As may be seen, the material can be compressed to 40% of its original volume by a pressure of 1000 bar, whereas compression to 50% requires only 70 bar. The maximum pressure required for compression is determined primarily by the required durability of the pellets.

The throughput of a given die ring is determined by the operation of the roller and the thickness of the new layer fed. The greater the diameter of the roller, the thicker the layer which can be drawn under it. Any superfluous material is pushed before the roller, the throughput does not increase in this case. The maximum effective layer

Fig. 2.36 Relationship between pressure and modulus of elasticity in pressing fodder flour

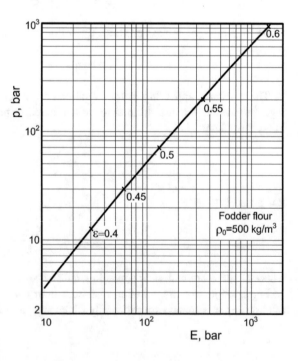

thickness depends on the friction coefficient between the bulk material and roller, and on the diameter of the roller (Dolgov and Vasilyev 1967)

$$y_0 = R - \sqrt{R^2 - 2r \cdot (1 - \cos \psi_0) \cdot (R - r)} \tag{2.72}$$

where

R—is the inside radius of the die ring,
r—is the radius of the roller,
ψ_0—is the friction angle.

The maximum angle of contact φ_0 (see Fig. 2.34) equates the friction angle

$$\varphi_0 = \psi_0$$

For example, if the friction coefficient is 0.7, then $\psi_0 = 35°$ and $\varphi_0 = 35°$.
An inspection of Eq. (2.72) reveals that y_o is the highest if the second term under the square root is maximum. Taking the derivative of this term in respect to r and equating to zero, yields

$$y_0 = y_{0max} \quad \text{if} \quad r = \frac{R}{2}$$

and

$$y_{0max} = 2r \left(1 - \sqrt{\frac{1 + \cos \psi_0}{2}} \right)$$

Figure 2.37 shows the throughput for one revolution as a function of the thickness of the layer fed for various roller diameter (Schwanghart 1969). The curves become horizontal for layer thicknesses above a certain value, depending on the diameter

Fig. 2.37 Pellet throughput as a function of layer thickness for one revolution

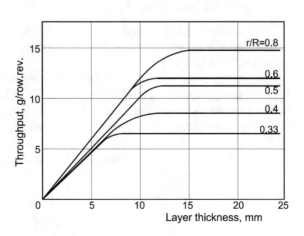

of the roller, i.e., the throughput fails to increase further. The throughput is also influenced by the friction coefficient between the roller and the material. The higher the friction coefficient, the thicker the layer which can be drawn in by the roller, and the greater the throughput.

An other point of view for system optimization may be the load on the rollers related to the throughput. With increasing roller diameter the contact surface area increases and together with it also the force acting on the roller and frame. According to investigations, the pelleting capacity per 1.0 t load on the roller is greatest when the relative roller dimensions r/R is in the range 0.3–0.4, for r/R above 0.5, the pelleting capacity increases only slightly while the load in the roller is highly increased (Schwanghart 1969).

2.8 Operational Characteristics of Die Ring Presses

Die ring press with two rollers is widely used for pelleting biomass. The main operating parameters are the throughput, the driving moment, rotation speed and power requirement. Knowing the throughput and power consumption, the specific power consumption can easily be calculated which contributes to the economy of pellet production.

The throughput is determined by the geometry of ring die, number of rollers, material properties and rotation speed

$$Q = 2R\pi bh_0\gamma_0 zn60 \ \ \text{kg/h}$$

where

z—is the number of rollers, mostly two,
n—is the rotation speed, rpm,
b—is the width of die ring, m,
γ_0—is the initial bulk density, kg/m^3.

The rotation speed is limited to around 100 rpm, at higher rotation speeds the compaction effect is worsening and the rollers incline to push the new loose material carpet (bulldozing effect).

The driving torque on the shaft of the rollers is given by the summation of tangential force components along the contact surface, Fig. 2.38. Due to the special driving mechanism, the rollers accomplish a combined motion. The radial force component in relation to the driving shaft is given by the relation

$$\Delta F_r = \sigma \cdot b \cdot \cos \psi \cdot \Delta s \ \ and \ \ \Delta s = r \cdot \Delta\alpha$$

or

$$\Delta F_r = \sigma \cdot b \cdot r \cdot \cos \psi \cdot \Delta\alpha$$

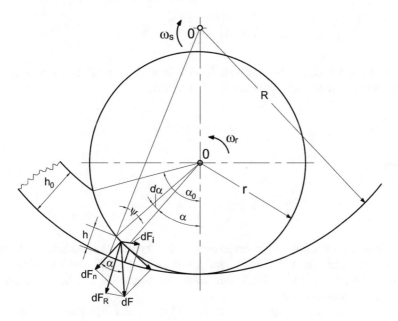

Fig. 2.38 Force relations under a pressing roller

With the assumption that $2r \cong R$,

$$\cos \psi = \sqrt{\frac{1 + \cos \alpha}{2}} \quad and \quad \alpha = 2\psi$$

The maximum value of ψ is around $17°$ and $\cos 17° = 0.9563$. Therefore, $\cos \psi$ may well be approximated with a constant value of 0.98. The tangential force component is calculated as

$$\Delta F_t = \Delta F_r \cdot \sin \psi$$

which should be integrated along the contact surface. The driving moment on the roller is

$$M = \sum \Delta F_t \cdot r$$

The power requirement on the driving shaft is given by

$$P = zM\omega \frac{R}{r} \quad and \quad \omega = \frac{n}{60} \cdot 2\pi = \frac{n}{9.55}$$

In order to perform calculations, the knowledge of pressure along the contact length is needed. The pressure-strain equation is used in the form

$$\sigma = A\left(\frac{\varepsilon}{1-\varepsilon}\right)^n \quad bar$$

and using beech chips, the constants have values of $A = 125$ and $n = 1.5$. The instantaneous strain ε at a given angle α is calculated as

$$\varepsilon = \frac{h_0 - h}{h_0 + h_g}$$

where

$$h = 2r\left(1 - \sqrt{\frac{1 + \cos\alpha}{2}}\right)$$

Substituting the friction coefficient as $\tan\alpha = \mu$, we obtain the maximum allowable material thickness h_0. For example, for $\mu = 0.7$ and $\alpha = 35°$, and r = 10 cm, the maximum thickness is $h_0 = 0.9257$ cm. the gap between roller and die ring is around $h_g = 1$ mm or somewhat less.

The above method requires numerical calculation which is, however, not complicated at all and can be accomplished with a better programmable calculator. As an example, with R = 20 cm, r = 10 cm, b = 5 cm, friction coefficient of beech chips μ = 0.7, contact angle $\alpha_0 = 35°$, maximum pressure 1400 bar, number of rollers is z = 2, rotation speed is 100 rpm, the initial bulk density is $\gamma_0 = 220$ kg/m^3, gap between roller and die ring is 0.8 mm, the expected throughput is 1536 kg/h. Integration of tangential forces gives a driving moment on the roller $M = 1462$ Nm and the net power consumption amounts to 61.24 kW which corresponds to a specific power consumption of 39.87 kWh/t. The energetic efficiency of roller presses is around 65–70% and the expected actual specific energy consumption is 51–61 kWh/t. These values agree well with measured values in the practice (see Fig. 3.14).

Table 2.3 shows the variation of strain, stress and tangential force (N per one degree) as a function of contact angle. The pressing of material into the boreholes begins at an angle of 10.7°. Thereafter the pressure remains more or less constant.

2.9 Effects of Various Parameters on the Pelleting Process

The density and durability of pellets made under a given pressure depend on numerous factors, the most important of which are the materials structure, temperature, initial volumetric weight and moisture content, the velocity of loading and the duration for which pressure is maintained.

With increasing moisture content, many agricultural materials assume plastic properties, facilitating compression. Thus a given volumetric weight may be attained with a lower pressure if the moisture content is higher. However, a given volumetric

Table 2.3 Strain, stress and tangential force on the roller

α	35	30	25	20	15	10.7	8	5	3	0
ε	0	0.24	0.45	0.62	0.75	0.83	0.83	0.83	0.83	0.83
σ, bar	0	23	92	258	653	1400	1400	1400	1400	1400
ΔF_t, N/1°	0	50	170	389	728	1124	835	522	313	0

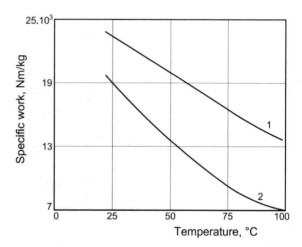

Fig. 2.39 Specific energy requirements for pelleting straw and hay as a function of temperature (1) straw, 12.6%, (2) hay, 9.6%

weight relative to the dry content can generally be attained only by higher pressure if the moisture content is higher.

The compressibility increases also with heating for many materials, permitting the same volumetric weight to be attained by a lower pressure. This means that the specific work requirements of pelleting may be reduced by preheating the material. Figure 2.39 shows the variation of the specific work required to pellet straw and hay (pellet diameter 78 mm) as a function of temperature. In pressing farinaceous materials it has been found that the shear strength (characterizing the stability) is highest in relation to the pelleting pressure for the temperature range 60–80 °C. This means that the energy consumption needed to attain greatest stability is also optimal in this temperature range.

The final pellet density can considerably be increased by using higher temperatures. Figure 2.40 shows the effect of pressing temperature on the density of pellet

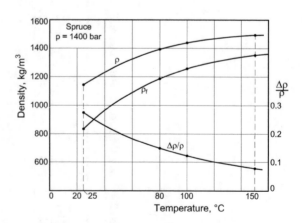

Fig. 2.40 Effect of temperature on the density of pellet under pressure and after recovery. Spruce, particle size 0.2–0.5 mm

under pressure and after recovery. The relative recovery decreases indicating the increasing plasticity of the material as the temperature increases (Kocsis 2015).

On increasing the length of the boreholes the energy requirements increase and the pellets produced are also stronger. For given dimensions and a given number of revolutions, increase of throughput lowers the strength of pellets.

2.10 Similarity Equation of the Energy of Compaction

During the compaction process the density and the modulus of elasticity of the material rapidly increase. Wood materials generally show non-linear viscoelastic-plastic behavior and, therefore, the pressure-deformation relationship is dependent on the loading velocity and on the time during which the material is subjected to constant deformation or load. The energy requirement of pellet production depends on many influencing factors and in such cases the use of dimensionless numbers in the form of similarity equation facilitates the processing of experimental results and the obtained similarity relationship has a more general validity for the users. Various fractions of different wood species were used in these experiments and the pressure, pellet diameter, temperature were also varied. The proposed similarity equation shows a good correlation with the experimental results.

The aim of present work is to develop and validate a relationship which takes into account the pressure, the final density of the pellet, pellet diameter, the effect of temperature and material properties. For this purpose the application of similarity equation (Buckingham 1914) seems to be the most promising method, similarly to those successful applications made in the last hundred years on the field of heat transfer and fluid flow problems.

The quality of the pellet meets the European norms for (EN 14961-12011). The optimum diameter (6 mm) and length (20–30 mm) of the pellets are fixed in this norm. We constructed press heads, that had diameter of 6, 8 and 16 mm and pressing channel length was 115 mm, this parameters are the same as the industry norm. In measurements we used a universal testing machine (Instron) (Fig. 2.41). The pressing velocity was 10 mm/min, which is the same as the industry norm. We used pressure parameters between 100 and 140 MPa, but also for special purposes between 30 and 300 MPa. We controlled and measured the temperature of the press head with a control system. Chips and sawdust are produced from spruce (*Picea abies*) and black locust (*Robinia pseudoacacia*) with 10–12% moisture content. In the final measuring process we changed the moisture content of the raw material between 5 and 15%. After the post chipping the raw material was put through a sieve vibrating 1.5 mm amplitude for 10 min. This produced 4 different particle sizes: 0.063–0.2 mm, 0.2–0.5 mm, 0.5–0.8 mm and 0.8–1.0 mm. We demonstrate the cumulative curve (spruce and black locust chip) using the lognormal integral distribution on a probability net. Average values were calculated from sieve analysis (Fig. 2.42 broken line).

(a)

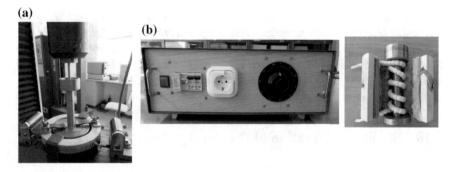

(b)

Fig. 2.41 Pressing cylinder and the temperature control of measurement system

Fig. 2.42 The particle size distribution on probability net

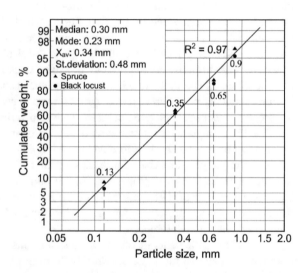

The predominant fractions are lying between 0.2 and 0.5 mm. The characteristic distribution parameters are the following: *Median* = 0.3 mm, *Mode* = 0.23 mm, geometric mean X_{av} = 0.34 mm and the standard deviation σ = 0.48 mm. The anatomical properties (soft or hardwood) influence the elastic- plastic behavior of the compressed material, especially the rebound of the material due to elastic recovery, after the pellet was pushed out from the channel. The increase of pellet diameter after recovery is a measure of the stored elastic stresses in the compressed state.

2.10.1   *Theoretical Considerations*

Biological materials have very complicated material laws and, with higher volume changes as is the pelleting process, the stress-strain relationship is always highly non-linear. Therefore pure mathematical methods for describing the compaction process in all its details are hardly available today. A more practical and reliable approach is to perform carefully designed experimental measurements and to process these results in such a way that the obtained relationship would be valid with so few constraints as possible. In order to extend the validity of a relationship for different materials, the inclusion of proper material properties seems to be indispensable.

The particles in the ring die channels are loaded by compressive forces and, therefore, as a first approximation, the material property may be characterized by the compressive strength of the material in question The compaction process of wood chips has the following main influencing factors material property of the wood species, average particle diameter, compaction pressure, temperature of the material, diameter of the pellet, final density of the pellet and the total specific work of the compaction. There are further influencing factors with a limited variability range due to the process itself. For instance, the loading speed always has a definite influence on the compaction process, but in real pelleting machines the loading speed cannot be varied as a process parameter. The moisture content is also an important process parameter, but its range is also limited (optimum 10–13%) due to the durability requirement of the pellet (Kocsis 2015; Kocsis and Csanády 2015).

Keeping in mind the above statements, the following formal functional relationship with seven variables may be written

$$W = f(p, \gamma, d, \sigma_C, \vartheta, \vartheta_0) \tag{2.73}$$

where

p—is the pressure, N/m^2,
W—is the total specific energy of compaction, Nm/m^3,
γ—is the final specific weight of the pellet, N/m^3,
d—is the diameter of the pellet, m,
σ_c—is the compressive strength of the wood, N/m^2,
ϑ, ϑ_0—are the pellet temperature and a reference temperature, here taken as $\vartheta_0 = 25\,°C$ (room temperature).

In order to derive the proper functional form of Eq. (2.73), the standard dimensional analysis method was used (Buckingham 1914; Langhaar 1951). It yields the following dimensionless numbers

$$\pi_1 = \frac{W}{p}; \quad \pi_2 = \frac{\gamma \cdot d}{\sigma_C}; \quad \pi_3 = \frac{\vartheta}{\vartheta_0}$$

and according to Buckingham's theorem, we may further write

$$\frac{W}{p} = f(\frac{\gamma \cdot d}{\sigma_C}, \frac{\vartheta}{\vartheta_0})$$

or

$$\frac{W}{p} = const \cdot \left(\frac{\gamma \cdot d}{\sigma_C}\right)^n \cdot \left(\frac{\vartheta}{\vartheta_0}\right)^m \tag{2.74}$$

where the constant and the exponents n and m should be determined experimentally.

In the above equation the dependent variable is the specific energy consumption W, and therefore, the dimensionless number π_1 containing the specific energy is placed on the left side of Eq. (2.74). The other two numbers containing only independent variables are on the right side of Eq. (2.74). The total work of compaction is determined by the pure compression work and by the work done during pushing out the pellet from the boreholes. The pure compression work is given by integration of the pressure along the displacement. The pressure—strain relationship is strongly non-linear and can be given according to Eq. (2.51) (Sitkei 1981).

$$p = A\left(\frac{\varepsilon}{1 - \varepsilon}\right)^n$$

where

A—is the material dependent constant,
ε—is the strain,
n—is the exponent.

The second main part of the total compaction work is the work done during the pushing-out of the pellet from the boreholes. This work depends on the friction coefficient between the pellet and the channel wall, the length of the channel and the elastic-plastic behavior of the material at the given pressure. This work should also be determined experimentally.

2.10.2   Experimental Results and Similarity Equation

Figure 2.43 shows the total specific work for spruce chips of different pellet diameters using 140 MPa pressure. With increasing diameters the relative contribution of wall friction decreases and therefore the energy requirement also decreases.

Heating pellets at different temperatures the viscoelastic properties of wood materials change considerably. Figure 2.44 shows the effect of pellet temperature on the total specific compaction energy.

The beneficial effect of heating is more efficient, however only up to 100–110 °C. A further increase of temperature reduces the energy requirement to a lesser extent. A possible explanation for this may be the reduction in friction coefficient between particles and in the virtual viscosity of the material as the temperature increases. Due to the limiting effect of solid wood density on the compaction process, the compaction is an asymptotic phenomenon with decreasing effectivity toward high densities. An inspection of Figs. 2.44 and 2.46 clearly shows that the pellet density over 100 °C is not far from the solid wood density (~1580 kg/m³). It should also

Fig. 2.43 The change in
specific work depending on
the diameter of spruce pellets
at ambient temperature

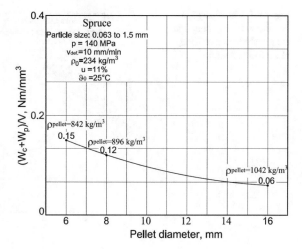

Fig. 2.44 The change of
total specific work depending
on pellet temperature using
spruce chips

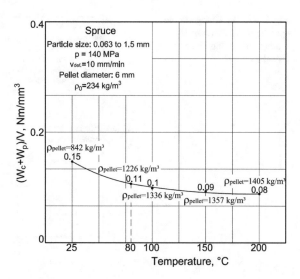

be remembered that the pellet density is always less than the compaction density at
maximum pressure due to the rebound.

Similar measurement results are depicted in Figs. 2.45 and 2.46 for black locust.

It is interesting to note that the total for black locust only slightly differ from
those of spruce although the components are not the same.

In order to calculate the dimensionless numbers, the compressive strength of the
wood species used is needed. In general this value varies for spruce in the range of
45–55 N/mm² and for black locust 60–65 N/mm² (Molnár 1999). It should be noted,
however, that the strength for small particles is not exactly the same as for the solid

Fig. 2.45 The change in specific work as a function the punch diameter using black locust samples

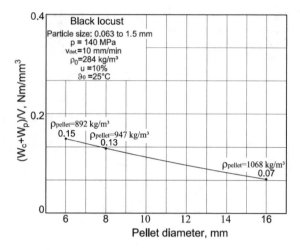

Fig. 2.46 The change in the specific work as a function of temperature using black locust chips

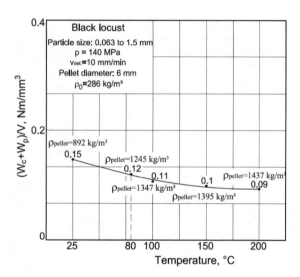

wood. Therefore, a slight correction might be required. The measured and calculated values for spruce and black locust are summarized in Table 2.4.

Plotting the dimensionless numbers has revealed that the selection of compressive strength values of 55 and 65 N/mm² for spruce and black locust respectively is appropriate and all measurement points are on the same line as shown in Fig. 2.47.

We investigated the influence of temperature on specific energy required to produce pellets from spruce and black locust chips at pressures between 100 and 140 MPa. Table 2.5 shows the results. Figure 2.48 summarizes the results in a dimensionless graph.

The exponent m in Eq. (2.74) should be determined so that the measured points for different pellet temperatures fit the same line properly as given in Fig. 2.47. Taking m

Table 2.4 Summary of research findings for pellets in the 100 and 140 MPa pressure range with 0.063–1.0 mm particle size at 25 °C

Species	Pressure	Specific energy	Specific energy	W/p	Pellet specific gravity	Pellet diameter	σ_C	$\gamma d/\sigma_C$
	p	W	W		γ	d		
	N/m^2	Nm/mm^3	Nm/m^3		N/m^3	m	N/m^2	
Black locust	1.4×10^8	0.15	1.5×10^8	1.07	8920	0.006	65×10^6	8.93×10^{-7}
	1.4×10^8	0.13	1.3×10^8	0.93	9470	0.008	65×10^6	11.65×10^{-7}
	1.4×10^8	0.07	0.7×10^8	0.50	10,680	0.016	65×10^6	26.28×10^{-7}
	1.0×10^8	0.12	1.2×10^8	1.20	8130	0.006	65×10^6	7.50×10^{-7}
	1.0×10^8	0.10	1.0×10^8	1.00	8340	0.008	65×10^6	10.26×10^{-7}
	1.0×10^8	0.05	0.5×10^8	0.50	9720	0.016	65×10^6	23.92×10^{-7}
Spruce	1.4×10^8	0.15	1.5×10^8	1.07	8420	0.006	55×10^6	9.18×10^{-7}
	1.4×10^8	0.12	1.2×10^8	0.86	8960	0.008	55×10^6	13.03×10^{-7}
	1.4×10^8	0.06	0.6×10^8	0.43	10,420	0.016	55×10^6	30.31×10^{-7}
	1.0×10^8	0.12	1.2×10^8	1.20	7030	0.006	55×10^6	7.67×10^{-7}
	1.0×10^8	0.09	0.9×10^8	0.90	7500	0.008	55×10^6	10.90×10^{-7}
	1.0×10^8	0.05	0.5×10^8	0.50	8450	0.016	55×10^6	24.58×10^{-7}

$= 0.15$, all measurement points, including also those for heated pellets fit the straight line as shown in Fig. 2.48. Points for elevated temperatures are marked with x.

The scattering zone of measurement points is fully acceptable and it corresponds to a good engineering accuracy. The calculated correlation coefficient is around 0.97.

In the following the effect of particle size on the compression work is examined. For these experiments we have used three fractions of chips for both spruce and black locust. The fractions have the following particle ranges: 0.063–0.2 mm, 0.2–0.5 mm and 0.8–1.0 mm, which were demonstrated in Fig. 2.42 The measurement and calculated results are plotted in Figs. 2.49 and 2.50. It is clearly seen that in both cases the energy requirement changes very little, although the slight decrease for both wood species has the same tendency. If we use now these additional data in the similarity

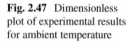

Fig. 2.47 Dimensionless plot of experimental results for ambient temperature

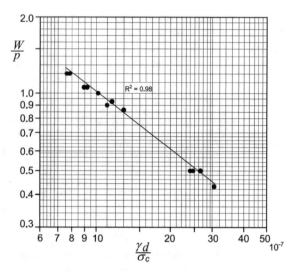

equation without any correction and plot quite similarly to Figs. 2.47 and 2.48, we obtain Fig. 2.51 including all measurement points.

The scattering of data points is not much higher and the correlation coefficient in this case is as high as 0.94. This is due to the fact that the middle fraction from the three was used for construction of Figs. 2.47 and 2.48 and therefore, the points of the other two fractions are placed on the opposite sides of the resultant straight line. The final similarity equation is given in the following form (Table 2.6)

$$\frac{W}{p} = C \cdot \left(\frac{\gamma \cdot d}{\sigma_C}\right)^{-0.75} \cdot \left(\frac{\vartheta}{\vartheta_0}\right)^{0.15} \tag{2.75}$$

where the constant C has the value of 3.12×10^{-5}.

Finally it should be noted that in these experiments a plunger with a bottom face was used. In practice, however, due to the continuous operation requirement, the chips will be pressed into the boreholes of a die ring and the counter-force is assured by friction forces on the channel wall. Therefore, the push-out force is more or less the same as the maximum compression force. This means that under real conditions somewhat higher specific energy is required.

2.11 Wall Friction and Channel Length

The ratio of the channel diameter (D) and the channel length (L) has significant influence on the pellet density due to the wall friction which is sometimes referred to as an odometer problem. In rolling presses, the counter pressure is ensured by wall friction forces which depend on the channel length. To our best knowledge

Table 2.5 Summary of measured and calculated results at 100 and 140 MPa, 0.063–1.0 mm particle size with temperature effect

Species	Pressure p N/m^2	Specific energy W Nm/mm^3	Specific energy W Nm/m^3	W/p	Pellet specific gravity γ N/m^3	Pellet diameter d m	σ_C N/m^2	$\left(\dfrac{\gamma d}{\sigma_C}\right)\left(\dfrac{\vartheta}{\vartheta_0}\right)^{0.15}$
Black locust	1.4×10^8	0.12	1.2×10^8	0.86	12,450	0.006	65×10^6	13.67×10^{-7}
	1.4×10^8	0.11	1.1×10^8	0.79	13,470	0.006	65×10^6	15.29×10^{-7}
	1.4×10^8	0.10	1.0×10^8	0.71	13,950	0.006	65×10^6	16.74×10^{-7}
	1.4×10^8	0.09	0.9×10^8	0.64	14,370	0.006	65×10^6	18.04×10^{-7}
	1.0×10^8	0.09	0.9×10^8	0.90	11,200	0.006	65×10^6	12.30×10^{-7}
	1.0×10^8	0.08	0.8×10^8	0.80	11,960	0.006	65×10^6	13.58×10^{-7}
	1.0×10^8	0.07	0.7×10^8	0.70	12,870	0.006	65×10^6	15.44×10^{-7}
	1.0×10^8	0.07	0.7×10^8	0.70	13,110	0.006	65×10^6	16.46×10^{-7}
Spruce	1.4×10^8	0.11	1.1×10^8	0.79	12,260	0.006	55×10^6	15.91×10^{-7}
	1.4×10^8	0.10	1.0×10^8	0.71	13,360	0.006	55×10^6	17.92×10^{-7}
	1.4×10^8	0.09	0.9×10^8	0.64	13,570	0.006	55×10^6	19.24×10^{-7}
	1.4×10^8	0.08	0.8×10^8	0.57	14,050	0.006	55×10^6	20.84×10^{-7}
	1.0×10^8	0.09	0.9×10^8	0.90	10,520	0.006	55×10^6	13.65×10^{-7}
	1.0×10^8	0.08	0.8×10^8	0.80	11,100	0.006	55×10^6	14.89×10^{-7}
	1.0×10^8	0.07	0.7×10^8	0.70	12,360	0.006	55×10^6	17.53×10^{-7}
	1.0×10^8	0.06	0.6×10^8	0.60	12,860	0.006	55×10^6	19.08×10^{-7}

Fig. 2.48 Measurement points including also those for heated pellets on double logarithmic scale

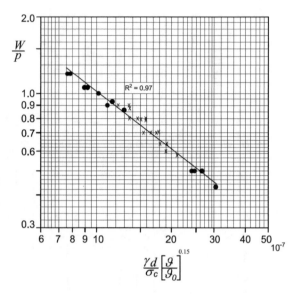

Fig. 2.49 The change in the specific work as a function of particle size using black locust

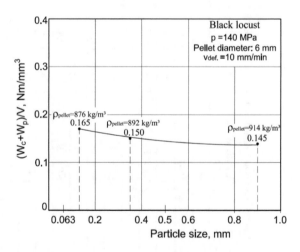

this problem has not yet been treated in details and, therefore, a theoretical and experimental investigation was undertaken to derive a generally valid relationship so describe the above phenomenon. The obtained relationship in dimensionless form is suitable to determine the necessary channel length for a given channel diameter for different pellet densities or required maximum pressure using chips of two wood species. The obtained results are in agreement with real channel diameter/length ratios used in practice.

Using die-ring pelleting machines, where the loose chip material is compressed by a roller and pressed into the boreholes of the ring, a common problem is the improper selection of the borehole diameter and length ratio for a given wood chip.

Fig. 2.50 The change in the specific work as a function of particle size using spruce samples

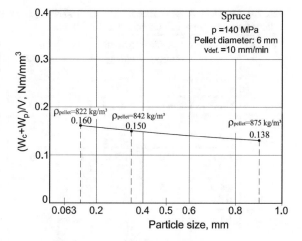

Fig. 2.51 The final similarity plot of all measurement results

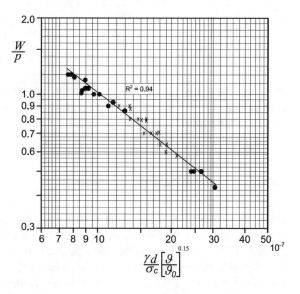

Various wood species may have quite different physical properties influencing the friction coefficient between the channel wall and the pressed material. As a consequence, a given die-ring can produce different pellet densities depending on the chip properties. In practice, the pellet manufacturers may incorrectly select the die-ring process parameters (wood species, particle size distribution, moisture content, temperature and the required compression pressure). Therefore, the results may be a low quality products or increased energy consumption. The compaction process in a press channel is a very complicated one due to the varying conditions along the press channel. The modulus of elasticity, the Poisson's ratio and the friction coefficient on the wall continuously vary along the length of the channel which make a pure

Table 2.6 Summary of measurement results at 140 MPa pressure and with three different particle fractions 0.063–0.2 mm, 0.2–0.5 mm and 0.8–1.0 mm

Species	Pressure	Specific energy	Specific energy	W/p	Pellet specific gravity	Pellet diameter	σ_C	$\gamma d/\sigma_C$
	p	W	W		γ	d		
	N/m^2	Nm/mm^3	Nm/m^3		N/m^3	m	N/m^2	
Black locust	1.4×10^8	0.165	1.65×10^8	1.18	8760	0.006	65×10^6	8.08×10^{-7}
	1.4×10^8	0.150	1.5×10^8	1.07	8920	0.006	65×10^6	8.93×10^{-7}
	1.4×10^8	0.145	1.45×10^8	1.04	9140	0.006	65×10^6	8.74×10^{-7}
Spruce	1.4×10^8	0.160	1.6×10^8	1.14	8220	0.006	55×10^6	8.96×10^{-7}
	1.4×10^8	0.150	1.5×10^8	1.07	8420	0.006	55×10^6	9.18×10^{-7}
	1.4×10^8	0.138	1.38×10^8	0.99	8750	0.006	55×10^6	9.54×10^{-7}

theoretical treatment impossible. In the following, a brief discussion of the problem is given which facilitates the proper processing of the experimental results and to get a useful relationship for practical purposes.

2.11.1   Theoretical Considerations

Due to the wall friction, the pressure along the length of press channel is not constant. Poisson's ratio and the friction coefficient also vary as a function of length (Biot 1954; Findley 1976; Heiko et al. 2005; Hofko 2006). One of the main influencing factors, the length/diameter ratio (*L/D*), plays a distinct role determining the influence of wall friction on the density distribution in the compaction channel. The pressure distribution in the space before a compressing piston can be given by Eq. (2.65) (Sacht 1967)

$$p_x = p_0 e^{-kx}$$

where

p_0—is the pressure exerted by the piston, N/mm^2,
p_x—is the pressure at a distance x from the piston, N/mm^2.

and

$$k = \frac{4}{D} \mu \frac{v}{1-v}$$

with

μ—is the friction coefficient,
v—is the Poisson's ratio (0.35–0.4).

The constant k in Eq. (2.66) contains the ratio of the wall surface area (F_{wall}) to the volume (V) which is

$$k' = \frac{F_{wall}}{V} = \frac{D \cdot \pi \cdot L}{\frac{D^2 \pi}{4} \cdot L} = \frac{4}{D}$$

In our experiments piston diameters of 6, 8 and 16 mm have been used. Taking the average values for $\mu = 0.6$ and $v = 0.4$, the constant k has values of 0.267, 0.2 and 0.1 cm^{-1} respectively. With decreasing k-values the pressure distribution will be more uniform in the press channel. The Poisson's ratio generally slightly increases with higher pressures. In the pellet range 0.35 or 0.4 is appropriate. The friction coefficient is influenced by the wood species, moisture content and pressure. An interesting observation for agricultural materials was that, applying high pressure, the water might be pressed out from the material reducing the friction coefficient considerably. The condition for water release is that the loading pressure is greater than the water potential (tension) of the material at a given moisture content. In the experiments an average moisture content of 10% was used. It corresponds to a pF-number of 6 or 1000 bar tension using the sorption isotherm and converted to tension. The maximum compression pressure was 1400 bar and, therefore, it may be assumed that some water is released and pressed out to the boundary of the particles. Furthermore, it may also be assumed, although never proved, that this water pressed to the particle boundaries plays an important role in producing a durable pellet by providing the necessary bonding forces.

2.11.2 An Approximate Modeling of the Pushing Force

As outlined above, the force required for pushing out the pellet from the channel originates from the side wall friction. This force formally can be simply calculated by the following equation

$$F_p = A \cdot p_w \cdot \mu \tag{2.76}$$

where

A—is the instantaneous contact area of the pellet with the channel wall, mm^2,
p_w—is the average side pressure, N/mm^2,
μ—is the friction coefficient.

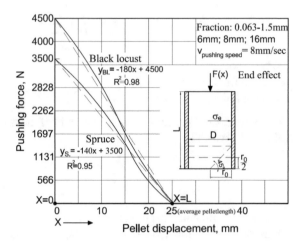

Fig. 2.52 The reduction of maximum pushing force as a function of free pellet length

The side pressure is developed by the stored elastic deformation of the material in the compressed state, and the magnitude of this deformation is the main problem.

Figure 2.52 shows the variation of pushing force determined experimentally as a function of displacement for pellets of 25 mm length. Black locust requires somewhat higher force compared to spruce. It is interesting to note that pellets with different diameters required practically the same force. Another important observation that the force varied nearly linearly as a function of displacement. Therefore, as a first approximation, in the following we calculate with a linear force function along the pellet length. The bottom end of the pellet leaving the channel will undergo an unloading process due to the free expanded part of the pellet (end effect in Fig. 2.52). For the sake of simplicity this effect will be taken into account with the assumption that the radial stress decreases, starting from the bottom edge of the channel, under 45° to its stationary value (Timoshenko and Woinowsky-Krieger 1957).

It is interesting to compare the maximum pushing force with the maximum compaction force F_c (1400 bar maximum pressure) for different pellet diameters and wood species (F_p = 3500 and 4500 N for spruce and black locust respectively). These F_p/F_c values are summarized in Table 2.7.

It is clearly seen that with increasing pellet diameters the share of the push out energy in the total energy of pelleting sharply decreases. As a pellet is being pushed-out, the contact area in Eq. (2.76) varies according to the following equation

$$A_x = D\pi (L_{pellet} - x) \tag{2.77}$$

Table 2.7 F_p/F_c force ratios for the investigated pellets in per cent

	$\Phi 6$ mm (%)	$\Phi 8$ mm (%)	$\Phi 16$ mm (%)
Spruce	88	50	12
Black locust	114	64	16

where

x—is the length of the pellet out of the channel, mm.

The instantaneous pushing force is

$$F_{(px)} = A_x \sigma_e \mu \tag{2.78}$$

where σ_e—is the stress due to the stored elastic deformation (lateral pressure), N/mm^2.

This stress is related to the elastic recovery of the pellet in the following manner

$$\sigma_e = \frac{E}{1 - \nu} \cdot \frac{\Delta r}{r_0} \tag{2.79}$$

where

Δr—is the increment of pellet radius due to elastic recovery, mm,
r_0—is the channel radius, mm,
E—is the stored true modulus of elasticity, (do not confuse it with the deformation modulus of the pellet in compressed state), N/mm^2.

As mentioned above, at the lowest end of the channel the stress σ_e declines (end effect) and this effect may be taken into account with the proper correction of the contact surface (see Fig. 2.52). The ratio of the effective area to the geometric contact area (surface reduction factor) is given by

$$\psi_{end} = \frac{D\pi (L_{pellet} - r_0/2)}{D\pi \cdot L} = \frac{L_{pellet} - r_0/2}{L} \tag{2.80}$$

where

L_{pellet}—is the pellet length, mm.

Keeping in mind the above equations, the instantaneous pushing force is

$$F_{(x)} = 2\pi \frac{E}{1 - \nu} \mu \Delta r (L_{pellet} - x) \psi_{end} \tag{2.81}$$

After pushing out the pellets, their diameters were measured. Because the increment Δr generally has values in the range of tenths of a mm, so a very accurate measurement is difficult. Average values are summarized in Table 2.8 for both wood species and for different pellet diameters.

Table 2.8 Increment of pellet diameters, Δr

	$\Phi 6$ mm	$\Phi 8$ mm	$\Phi 16$ mm
Spruce (mm)	0.25	0.3	0.52
Black locust (mm)	0.245	0.3	0.51

Table 2.9 Elastic modulus $E/(1-\nu)$ in the compressed pellet

	$\Phi6$ mm	$\Phi8$ mm	$\Phi16$ mm
Spruce (N/mm^2)	146	124.2	78.5
Black locust (N/mm^2)	221	188.7	119

Based on our measurements, no significant difference could be established between the two wood species. Sadly, friction coefficients for wood chips under high pressure conditions are not available at all. The obtained functional relationship Eq. (2.81) does not allow separating the effect of friction and lateral pressure on the pushing force. Nevertheless, a good estimate can be made in the following manner.

We possess detailed experimental results on friction properties of wood species for particle size distributions and surfaces (Varga 1983). Under low pressure the appropriate values are $\mu = 0.65$ and $\mu = 0.55$ for spruce and black locust respectively. If we do not change these values for high pressures then the true modulus of elasticity of the compressed pellet can be calculated. Knowing the pushing force, Eq. (2.81) is suitable to calculate $E/(1-\nu)$ values for different pellet diameters. They are summarized in Table 2.9.

It is probably that the friction coefficient under high pressure is somewhat less than the above figures. In this case the elastic module will be higher in the same ratio as the friction coefficients were lowered.

The actual deformation modulus of the pellet is naturally much higher. For comparison, an estimate can be done very easily. Using the nonlinear compaction equation, according to Eq. (2.51) for Spruce chip compaction ($D = 6$ mm, $p_c = 1400$ bar, $\varepsilon = 0.8$) the following equation is valid

$$p_c = A\left(\frac{\varepsilon}{1-\varepsilon}\right)^n$$

where $A = 8.75$ N/mm^2 and $n = 2$.
Derivation of this equation gives the deformation modulus as in Eq. (2.52)

$$E_d = \frac{\partial p_c}{\partial \varepsilon} = A \cdot n \left(\frac{\varepsilon}{1-\varepsilon}\right)^{n-1} /(1-\varepsilon)^2$$

Substituting the appropriate values, we get $E_d = 1750$ N/mm^2 which is one magnitude higher than the true elastic modulus in the compressed pellet. Using black locust chips, the required strain is somewhat less ($\varepsilon = 0.75$) and the deformation modulus is $E_d = 1590$ N/mm^2. The difference in the deformation may be explained by the difference in hardness of the individual particles.

2.11.3   Determination of Channel Length for a Given Pressure

Using die-ring pelleting machines with continuous operation, the prescribed maximum compaction pressure (or force) is in equilibrium with friction forces acting on the channel surface. Using Eq. (2.81), the equilibrium equation has the form ($x = 0$)

$$p_c = \frac{F_c}{r_0^2 \pi} = \frac{2}{r_0} \frac{E}{1-\nu} \mu \frac{\Delta r}{r_0} L \psi_{end} \tag{2.82}$$

It seems to be advisable to transform the above equation into a dimensionless form

$$\frac{L}{D} = \frac{p_c}{4 \mu \psi_{end} \frac{E}{1-\nu} \cdot \frac{\Delta r}{r_0}} \tag{2.83}$$

where

$D = 2r_0$—diameter of the channel, mm,
L—is the channel length, mm,
E—is the true modulus of elasticity in the compressed pellet, N/mm^2.

Using the measurements results with the three channel diameters, the functional relationship between diameter and channel length is shown in Fig. 2.53, both for spruce and black locust, and at $p_c = 1400$ bar.

Spruce chips, for the same compaction pressure, require a slightly longer channel compared to black locust. Therefore, using a given die-ring, different compaction pressures may be obtained if chip composition varies. The more generally valid dimensionless plot of experimental results is given in Fig. 2.54, which is a direct proof of the validity of Eq. (2.83).

Equation (2.83) is implicit in respect to the channel diameter and, therefore, it does not show the true relationship between channel length and diameter. Numerical calculations show that this relationship with good approximation is a quadratic

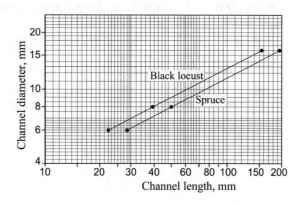

Fig. 2.53 Relationship between channel diameters and channel length ensuring the required compression pressure

Fig. 2.54 Representation of Eq. (2.83) based on experimental results

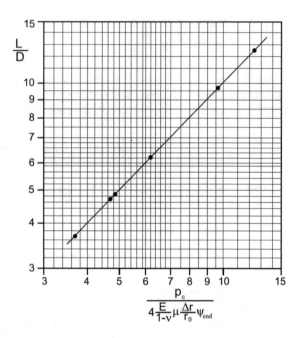

function in the form

$$L = const.D^2 \quad or \quad \frac{L}{D^2} = const. \tag{2.83a}$$

The *constant* is material depend and it amounts to 7.8 and 6.15 for spruce and black locust respectively, if cm dimension is used.

Obviously, for other wood species and chip particle distributions the experimental determination of E and Δr values is necessary. As Table 2.9 and Fig. 2.53 clearly suggest, the rotating die ring pelleting method allows to the use of channel diameters in a very limited range (5–7 mm dia.) due to the rapid increase in the necessary channel length with increasing diameters.

From a practical point of view, the pelletizing pressure and the press channel length depend on the density of the feedstock.

Practical observations and experimental results have also shown that the higher the density of the feedstock, the smaller the pressure and the required pressing channel length to obtain a given density. We have seen (Fig. 2.53) that for compression pressures with a low density (spruce) about 30% longer channels are required than for black locust. However, at constant pressure, the density of the spruce pellets is less than that of the black locust pellets, as shown in Figs. 2.43, 2.44, 2.45 and 2.46. This means that in order to reach the same pellet density with spruce, a channel needs to be 25% longer.

2.12 Optimum Selection of Operational Parameters

In order to get durable pellets with minimum investment, labour and energy costs, the operational parameters, to the given raw material properties, should be selected optimally. The main operational parameters are the maximum pressure, the channel length and diameter which ensure the required final density for durable pellets. In this respect we should strictly distinguish pellet density of maximum pressure and final density after recovery.

The final density of pellets is obtained as a result of subsequent calculations in several steps and there is no explicit relationship between final density and maximum pressure, and channel sizes. Using spruce and black locust chips, the following explicit equation is found

$$p = const. \frac{\rho_f^n}{(L/D^2)^m} \quad bar \tag{2.84}$$

where the constant and the exponents n and m are material dependent. Due to the fact that the relative recovery along the compression curve is changing (see Fig. 2.15), therefore the pressure-final density curve is distorted compared to the pressure-density curve. As a consequence, the exponent n is not constant for a wider pressure range. Fortunately, the feasible pressure range lies between 1000 bar and 1400 bar and for this range an average exponent can be adapted. The exponent m has a value of 0.29 while for exponent n values of 2.4 and 1.7 are obtained for black locust and spruce respectively.

Further empirical equations have been found to estimate the effect of operational parameters on the final pellet density. Using spruce chips, the final pellet density may be estimated with the following equation

$$\rho_f = 16.3 \cdot p^{0.56} \cdot D^{0.2} \left(\frac{\vartheta}{\vartheta_0} \right)^{0.36} \quad kg/m^3 \tag{2.85}$$

and for black locust

$$\rho_f = 63 \cdot p^{0.38} \cdot D^{0.2} \left(\frac{\vartheta}{\vartheta_0} \right)^{0.28} \quad kg/m^3 \tag{2.86}$$

where

D—is the channel diameter in cm,
P—is the pressure in bar,
ϑ_0—is the reference temperature taken as 25 °C.

The increase of channel diameter slightly increases the expected final density due to the decreasing friction forces. Using higher pellet temperatures, the final density increases considerably. The exponent of pressure depends on the hardness and mean diameter of particles.

Table 2.10 Range of operational parameters	Geometric mean of particles	0.5–1.0 mm
	Moisture content	10–12%
	Channel diameter	6–8 mm
	Channel relative size (L/D^2)	6–8 1/cm
	Pressure range	1000–1400 bar
	Final density	900–1100 kg/m^3

In the pellet production the feasible range of operational parameters generally is rather narrow which facilitates the optimum selection of their values. The range of main operational parameters may be given as shown in Table 2.10.

In the high pressure range over 1000 bar, it may be supposed that the final density of pellet approximately amounts to 82–85% of the maximum density before recovery.

As Table 2.10 shows, the narrow range of operational parameters does not make necessary a common mathematical optimization procedure. The economic production of pellets much more depends on the organization of the whole production system including raw material supply, a constant material flow in the subsequent preparation phases, maintenance of machines etc.

2.13 Influence of Moisture Content

It is well-known from the practice that the deviation of moisture content upwards or downwards in relation to the common values (10–13%) will highly decrease the mechanical stability of pellets, i.e., the binding forces among the particles are sharply decreased. The detrimental effect of higher moisture contents can easily be explained by the fact that much water in the material hinders the compaction of the material. It is not the case, with decreasing water content. Therefore, this latter case requires explanation. It is generally assumed that the binding forces among the particles originate from the lignin which is one of the main constituents of wood (Stamm 1964; Demirbas 2001; Andrew 2004; Anon 2009; Dmitry et al. 2013). Lignin is known to be a binding material on the outer surface of the cell walls which support the above assumption (Karwandy 2007; Escort 2009). But it is not fully clear why a small decrease in moisture content causes a rapid deterioration of binding forces among the particles and with this the loss of durability. It is therefore reasonable to suppose that water itself may play also a given role in the development of binding forces.

It is also known that molecular forces exert considerable action on water molecules contacting a surface in one or several layers. The highest force and adhesion will be exerted on the first layer of water molecules which may contribute to the binding forces considerably (Brunauer 1938; Hailwood and Horrobin 1946; Dent 1977; Pizzi et al. 1987). A layer of water molecules can appear on the surface of particles only, if the pressure used is somewhat higher than the water potential (tension) of wood material at a given moisture content. Using the theory of wood-water relations, the water potential curve can be constructed and used to check the validity of the above assumption.

Wood materials tend to be in equilibrium with the surrounding environment which is changing continuously in its temperature and relative humidity. At equilibrium the partial vapour pressure above the surface is equal to the saturated vapour pressure corresponding to the given temperature.

As mentioned above, wood is a porous material having capillaries of different diameters. On a liquid-gas interface, due to intermolecular forces, there is surface tension. As a consequence, water rises in the capillary and over the capillary meniscus, depending on the radius of the capillary, the static pressure decreases which corresponds to a decreased relative humidity over the meniscus. This relationship is described by the Thompson equation as follows

$$\phi = \exp(\frac{2\sigma \rho_v}{p_v \rho_w r})$$ (2.87)

where

φ—is the relative humidity,
σ—is the surface tension of water (0.072 N/m),
p_v—is the vapor pressure,
ρ_v, ρ_w—is the density of vapor and water, respectively,
r—is the radius of capillary.

It is interesting to note that the sap in wood contains soluble nutrients which decrease the surface tension to about 0.05 N/m. As an example, a capillary radius r = 0.36 nm corresponds to φ = 5% and r = 100 nm to φ = 98%. It means that in the sorption process, capillaries with a radius over 100 nm cannot be filled with water. That means that the sorption isotherm does not characterize the entire moisture range of a saturated wood sample. An approximate distribution of pore radii for a pine wood is represented in Fig. 2.55 with the corresponding moisture content (Sitkei 1994). The water taken up in the sorption process is called the hygroscopic water and the corresponding moisture content is the fiber saturation point (Tiemann 1906).

Measurements on the distribution of micro capillaries in wood have shown that most of the capillaries are in the 0.5–1.0 nm range which corresponds to φ = 20% and appr. 5% moisture content (Stamm 1946). This low relative humidity means a high driving potential or tension which rapidly decreases with increasing capillary radii. This capillary tension is a real driving force which can be expressed in term of water potential as

$$\psi = \frac{2\sigma \cos \theta}{r} \ (Pa)$$ (2.88)

where

θ—is the contact angle to the wall of the capillary tube.

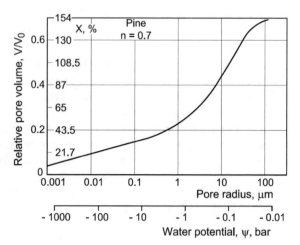

Fig. 2.55 Relative pore volume of pinewood as a function of pore radii. The corresponding theoretical water potential values are also given

Due to its relative humidity, the surrounding air has also a driving potential or tension given by the following equation

$$\psi_a = \frac{RT}{m_w} \ln \varphi \quad (\text{bar}) \tag{2.89}$$

where

m_w—is the molecular volume of water (18 cm^3/mol),
T—is the temperature (°K),
R—is the gas constant (82 bar cm^3/mol °K).

Figure 2.56 gives a relationship between the equilibrium moisture content and the relative humidity of air. Therefore, using Eq. (2.89), the driving potential of the air can be calculated for each equilibrium moisture content. As an example, the air with a relative humidity of 90% has a water potential or tension of app. −140 bar. Using

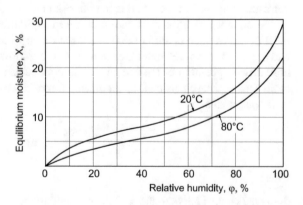

Fig. 2.56 The equilibrium moisture content of wood for two different temperatures

Fig. 2.57 The pF curve of
wood for two different
temperatures

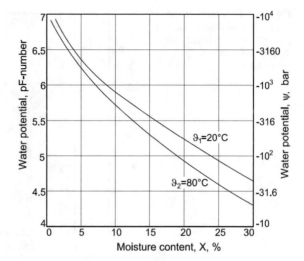

the sorption isotherm given in Fig. 2.56, the corresponding water potential curves
are depicted in Fig. 2.57.

The water potential may have different dimensions depending on the choice of
dimensions in Eq. (2.89). If we choice for $R = 8.314$ kJ/kmol °K and $m_w = 18$ kg/kmol,
then the water potential has the dimension kJ/kg. For example, taking 50% relative
humidity, it corresponds to nearly 100 kJ/kg and 1000 bar tension. Originally in soil
science, the pF-number was introduced to characterize water potential which is the
logarithm of the pressure height expressed in cm (Schofield 1935). For instance, pF
= 3 means 1000 cm = 1 bar tension.

For each given moisture content a series of measurements was conducted to find
the minimum pressure for producing stable pellets. The obtained minimum pressure
was then represented in the previously constructed water potential relationship as
shown in Fig. 2.58.

Because the water potential can be expressed in the common pressure unit, for
instance in bar, therefore the required minimum pressure can be directly compared
to tension holding the water in the material at the given moisture content.

The obtained results are highly interesting. The measurement points are system-
atically placed slightly above the water potential curves as a function of moisture
content. This means that the required compaction pressure should always be higher
than the water holding tension at a given moisture content. Increasing the moisture
content above 13%, the required pressure rapidly increases. In the presence of exces-
sive water, the compactibility of the material is worsening and a thicker water layer
among the particles decreases the effect of molecular forces.

The measurement results clearly indicate that some water may be pressed out from
the material to its surface giving an additional binding force among the particles. It is
well known in the soil science that a thin water layer (less than 5 or 6 molecule), on
the soil particles makes the water immobile, due to molecular forces and this water

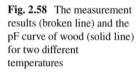

Fig. 2.58 The measurement results (broken line) and the pF curve of wood (solid line) for two different temperatures

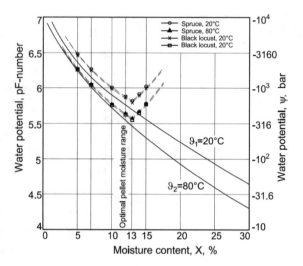

is not available for plants (wilting point). Similar observation was also presented studying water sorption energies (Pizzi et al. 1987). Therefore, it may be assumed that here also the thin water layer on the particle surfaces contributes to the binding forces.

The experimental results prove the practical experience that the optimum moisture content for making pellets is in a narrow range between 10 and 13%. Below and over this moisture range, the necessary pressure steeply increases, creates additional friction requiring much more compaction energy and to push the pellet out of the die.

It is interesting to note, that the required minimum pressures at 80 °C are somewhat nearer to the water potential curves than at 20 °C temperature. The absolute minimum pressure points for both cases are obtained at the same moisture content of 13%. According to these experiments, in the optimum range of moisture content a compaction pressure of 1000 bar produces stable pellets with sufficient safety.

These theoretical and experimental results prove that hygroscopic materials hold water by tension. From this statement it follows that the dewatering of a material is possible only by applying a pressure somewhat higher than the tension exerted by the material at a given moisture content. The same statement is valid for the pelleting process if the required minimum pressure which ensures a stable pellet is higher than the tension exerted by the material at given moisture content, then some portion of water will be pressed out of the chip particles to their surface. This thin layer of water may also contribute to the development of bonding forces.

Based on theoretical and experimental investigations the following main conclusions may be drawn

– The main influencing factor is the pellet diameter which fundamentally determines the role of wall frictionfriction forces in the total energy requirement.

- Heating the material reduces the required pressure by 30 and 35% at 100 °C degrees compared with 25 °C. Over 100 °C degrees the rate of reduction declines.
- The chip size distribution has some effects on the energy requirement but, it does not cause significant error.
- Using a similarity equation is also a powerful method to generalize experimental results for compaction processes. This is a simple and quick way to estimate the energy requirement.
- The length to diameter ratio of the press channel considerably influences the stress distribution in the pellet.
- In pellet making much energy is spent on the pushing the compressed pellet out of the channel. A new model is developed to determine the main relationships of this process.
- Using the experimental results with two different wood chips, the elastic recovery and the corresponding true modulus of elasticity were determined for several pellet diameters. This true modulus of elasticity is responsible for the lateral pressure on the channel wall and the associated friction forces.
- The maximum compaction pressure is in equilibrium with the friction forces acting on the channel surface. A dimensionless relationship is developed to determine the necessary channel length for a given compaction pressure and pellet diameter.
- The mechanical properties of a chip may considerably influence the appropriate channel length to ensure the required pressure and pellet quality.
- Moisture also has a definite role in the development of bonding forces in pellets,
- The relation of compaction pressure and water tension in the material can conveniently be followed using the water potential curves,
- It is essential to control the chip moisture content to minimize the energy consumption of pellet making.

Chapter 3
The Practice of Pellet Making

3.1 Introduction

This chapter discusses the important issues and problems of pellet making in practice. The individual working steps of pellet production such as material handling, chipping, drying, conditioning, pressing technologies, post-treatment and packing are introduced. The general layout of pellet making technologies is also treated. The energetic utilization of pellets is discussed in detail including the combustion process in furnaces, and a calculation method for design and process evaluation purposes. Pollutant emission problems are also discussed.

In order to complete the laboratory measurement results, practical observations for energy requirements for practical conditions are included. Finally, the economy of pellet production is analyzed in detail taking into account the expected investments, energy requirement of all possible operations, and a general cost calculation method is presented and illustrated with a model calculation.

General remarks
The efficiency and competitiveness of pellet manufacturers primarily depends on the production of an economical and high quality product. That is why mastering the technology of pellet production is crucial. In this chapter we summarize the most important practical experiences that can be used to manage high quality pellets. We also discuss the economic issues of the production of pellet and briquette manufacturing in detail which is the most important condition to be successful in the market.

There are a number of definitions for pellets and briquettes in publications and books, which, although well summarized, need some supplementation. Taking into account the research results of Chap. 2, the following general definition can be used to describe pellets and briquettes Pellets and briquettes are properly compressed particles of wood or agricultural materials using their own or mixed binder with appropriate moisture content and pressure. This definition applies primarily to pellets and briquettes made from wood and agricultural raw materials in the classic sense.

© Springer Nature Switzerland AG 2019
Z. Kocsis and E. Csanády, *Theory and Practice of Wood Pellet Production*,
https://doi.org/10.1007/978-3-030-26179-5_3

The pellets can be grouped as follows

– Feedstuff pellet. The most commonly known pellet species is rabbit and dog food. The feedstuff pellet has a diameter of 4–5 mm and a length of about 10 mm. The sugar industry calls them pellets, e.g. leached and then pelleted beet, which is sold as feedstuff.
– The "firing pellet" is common with a diameter of 6–8 mm and a length of 20–30 mm optimally. Depending on the material, we can talk about wood and agripellets. An agripellet contains various plant residues, while wood pellets contain sawdust and chips compacted properly.
– As long as the pressed biofuel is smaller than 50 mm in diameter, it is called a pellet while the bigger ones with a circular, rectangular or polygonal cross-sections are called briquettes. According to the raw material used, we can distinguish fuel briquettes made of brown coal, wood, bark, paper, straw, corn stalks or some other biomass.

Briquettes may have round or rectangular cross-sections. In the latter case, the edges have a higher surface/volume ratio which considerably facilitates the ignition process to start a steady burning speed. It is desirable that the briquette should not disintegrate during burning but, retain its original shape, would burn glowing.

3.2   Working Phases of Pellet Production

The pressing operation of pellets requires the presence of a properly prepared chipped material. We suppose that the collection, storage and transportation of raw material to the factory performed and the cost of the above operations is included in the delivery cost of the raw material. In general, the raw material delivered is debarked roundwood with a higher moisture content. In this case the main individual preparation processes necessary for pellet manufacturing are the following: primary cutting or chipping, drying, post-chipping and conditioning. The engineering solutions used for these operations depend on the starting raw material, especially on its size, moisture content and the desired pellet size and density.

3.2.1   Basic Material Properties

In principle, many different raw and waste materials, and residues may be used for pellet and briquette making. The different starting materials require different handling and preparation, and pressing technology due to their individual mechanical properties, size distribution, moisture content etc. The produced pellet or briquette may also have different properties, especially its heating value, ash content, melting temperature of ash, stoichiometric air consumption and the optimum air-fuel ratio.

Table 3.1 Composition and firing characteristics of pellet raw materials (Fenyvesi et al. 2008)

Name	N	C	S	H	O	Cl	Ash	Heating value (dry matter) MJ/kg
Poplar pellet	0.182	45.606	0.107	4.760	41.191	0.001	1.753	18.057
Beech pellet	0.182	45.838	0.063	5.621	39.912	0.001	0.584	19.016
Pine pellet	0.100	48.348	0.104	5.325	39.720	0.006	0.167	19.097
Willow pellet	0.781	44.638	0.118	4.665	37.988	0.030	4.320	18.101
Chinese reed	0.202	45.270	0.088	4.920	40.994	0.099	1.397	18.246
Sorghum pellet	0.742	42.149	0.190	4.772	37.725	0.500	7.503	16.757
Corn stalk	0.096	48.371	0.090	5.347	39.707	0.006	0.152	17.671
Hemp	0.555	43.572	0.151	4.792	40.768	0.072	3.410	17.229
Rape pellet	0.126	47.519	0.086	5.431	39.750	0.021	0.301	20.250
Straw pellet	0.663	40.114	0.187	3.907	37.179	0.900	7.800	16.665
Energy grass	1.502	46.447	0.205	6.012	31.424	0.120	5.500	17.226

Table 3.1 summarizes the composition and firing characteristics of possible pellet materials (Fenyvesi et al. 2008).

The type of combustion system, which can be used for burning of a given biomass, fundamentally depends on the firing characteristics and ash properties. Furthermore, the pollutants emitted such as carbon monoxide, hydrocarbons, nitrogen oxides and other compounds are also related to the composition of the raw material.

The inherent inorganic elements including silicon (Si), potassium (K), sodium (Na), sulphur (S), chlorine (Cl), phosphorus (P), calcium (Ca), magnesium (Mg) may be involved in reactions leading to ash fouling and shagging. Herbaceous fuels contain silicon and potassium which are their main ash-forming constituents. Their chlorine content is also high relative to other biomass fuels. These properties may cause severe ash deposition problems at high combustion temperatures. These ash problems come from the reactions of alkali with silica or sulphur forming alkali silicates and sulfates. The main source of alkali is potassium and the K_2O content in the ash is characteristic for a given fuel (Jenkins et al. 1998). In this respect, wood has far less ash content compared to high-ash straws, for example.

3.2.2 Primary Chipping

Pellet production for energy purposes started in the 1920s in North America mostly using sawdust. Sawdust is the most suitable material for direct pellet production without extensive preparations provided that the moisture content of the sawdust is suitable for pressing. In the ideal case, the sawdust was produced at the site of the pellet production and large storage capacity for sawdust was not required. A well organized material preparation is necessary in large scale production of pellets to ensure continuous operation. The first step is to cut the raw material into 10–20 mm pieces which are suitable for further preparation.

The primary rough chipping is usually done with disc, drum or screw chippers. Drum chippers are mainly used for crushing agricultural by-products. Hard metal knives are generally used and their life time depends on many factors such as the wood species (hardness or density, cutting direction), cutting parameters (tooth bite, depth of cut, and cutting speed) and on other parameters to a lesser extent. There is always a conflict between maximum production rate, maximum tool life and optimum production cost (Csanády et al. 2019). Fortunately, the specific cost function is flat as a function of cutting speed and, therefore, a good compromise can be made between optimum cost and maximum production rate.

3.2.3   Drying of Chips

Using debarked roundwood collected in the forest, its moisture content must be reduced to at least 10–13%. If it is possible, natural drying in the field may be an economic decision. The cutting energy slightly increases with decreasing moisture content but the saving on drying costs considerably surpasses the preceding cost increments. Today the most common practice is to artificially dry the moist raw material. The energy consumption of drying may be the half or somewhat more of the total energy consumed for the pellet production.

In the most cases, a drum or belt dryer is used. Rotating drum dryers operate with concurrent flow. The high temperature flue gas flows in the same direction as the material flow. In this case the cool material with its initial moisture content will be in contact with the high temperature flue gas which rapidly cools down to a lower temperature and does not cause the material to burn. In the rotating drum the material moves both vertically and horizontally which enhances the heat and mass transfer fundamentally. The drying process in a drum is controlled by the chip size, the temperature and velocity of the flue gas. Using smaller chips, the flowing flue gas may transport the dried particles due to their light weight.

In belt dryers, counterflow or multiple crossflow may be used with much lower gas temperatures compared to drum dryers. The speed of a conveyor belt can also be varied allowing a more precise control of final moisture content. Due to the lower gas temperature, wood or biomass firing can also be used for heat generation. To estimate the drying energy required for a given initial moisture content the specific energy of drying (MJ/t.% moisture) is needed. This value is around 55 MJ/t.% moisture. For more detailed calculations see Sects. 3.7 and 3.12.

3.2.4 Rechipping and Screening

The primary chipped and dried raw material is not suitable for pressing due to its rough size distribution. Depending on the selected pellet diameter, the necessary particle size varies between 0.2 and 2.0 mm. In order to achieve the required size distribution, generally a post-chipping operation is needed. For this purpose, a hammermill may be used which requires dried material. The comminution process is accomplished by hammers (swinging blades) rotating with high circumferential speed and hit the particles causing disintegration.

The energy for comminution highly depends on the brittleness of the material and, therefore, the use of dried material is best. The rotating hammers are surrounded by a screen which controls the exit size of the particles. Furthermore, many particles collide with the screen causing additional disintegration. The high speed rotation also generates an intensive airflow which transports the small particles out of the hammermill. The particles will be separated from the air in a cyclone. The fine particles remaining in the air after the cyclone can be separated in a mat-filter. For more details on the operation characteristics of hammermills and their energy consumption as a function final of chip size, see Sect. 3.10.

3.2.5 Conditioning

In the drying process, especially when using rotating high temperature drum driers, some overdrying always occurs. Below the optimum moisture content which varies in a narrow range-, between 10 and 13%, the required pressure for a durable pellet increases rapidly. Therefore, a moisture conditioning operation is worth including.

The best and shortest way for rewetting is to use saturated vapor. Theoretically, 10 kg vapor is needed to increase the moisture content of one metric ton material by one per cent. Because the equilibrium moisture content of wood in saturated vapor is around 19%, it is important to control the vapor dosing to the chip. Overdosing of vapor will lead to higher moisture content than is required for optimum pressing. From an economic point of view, it is advisable that the amount of rewetting is not higher than 2–4%. For this reason, the drying process should be controlled as much as possible. Cost estimates for conditioning are given in Sect. 3.12.

3.2.6 Pressing Technologies

The commercially available pressing machines operate in two different principles. The first type of machine exerts pressure in one or more direction onto the material. In these machines the pressure is exerted by a piston. It is interesting to note that multi-axial compression makes higher compaction compared to uni-axial compression.

This type of machine is capable to produce bigger compacted pieces in the briquett range.

Machines with a die ring are used in the pellet range with 6–10 mm diameters (Fig. 3.1).The bulk chip material fed is compressed by a suitably adjusted roller and pressed into the boreholes of the ring. The compressed material emerging from the boreholes is cut by knife on the outer side. The roller is set relative to the die ring with a given gap and a compressed layer formed on the running surface of the ring. Arriving at the minimum gap, the material is pressed into the boreholes.

The throughput of a given die ring is determined by the operation of the roller and the thickness of the new layer fed. The effective new layer thickness depends, on both the roller diameter and the friction coefficient between roller and chip material. The greater the diameter of the roller, the thicker the layer which can be drawn under it. Any superfluous material over a critical thickness is pushed before the roller and this material does not increase the throughput. The rotating motion, due to its radial acceleration, helps to maintain a stable material carpet on the inner surface of the die ring.

The increase of roller diameter is limited by the size of the ring. Furthermore, with increasing roller diameter, the contact surface area also increases which rapidly increases the force acting on the roller and frame. The optimum ratio of roller and ring diameters lie in the range of 0.3 and 0.4 (Schwanghart 1969). An other version of the above pelleting principle is the use of a round flat die with pressing rollers, (Fig. 3.2). In this design the boreholes are arranged in axial direction in several rows. In this design more rollers may be used which may increase the throughput. On the other hand, the compaction mechanism of the rollers is more complicated due to the circular path motion and different slips along the roller width.

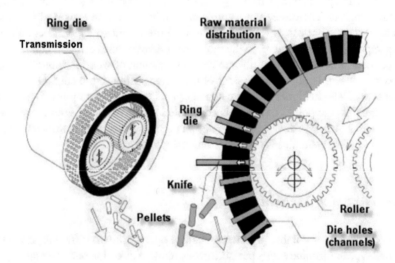

Fig. 3.1 The ring die pellet mill

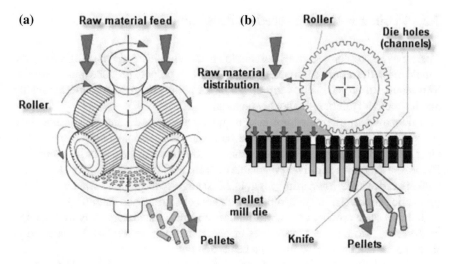

Fig. 3.2 The flat die pellet mill

The size of the boreholes, diameter and length, depends on the material properties and the required pellet density. Table 3.2 shows recommended channel lengths for a pellet diameter of 6 mm using different mixture ratios for soft- and hardwoods (Burján 2009).

It should be stressed, that the wood species and the required density may modify the channel length values given in Table 3.1 (see Sect. 2.11.3).

Table 3.2 Selection of channel length for 6 mm pellet diameter using different soft- and hardwoods ratios

Wood species mix ratio		Die parameters	
Hardwood, %	Softwood, %	Channel length, mm	Die type
100–90	0–10	14	A
90–80	10–20	17	B
80–70	20–30	21	C
70–60	30–40	24	D
60–50	40–50	26	E
50–40	50–60	28	F
40–30	60–70	30	G
30–20	70–80	32	H
20–10	80–90	34	I
10–0	90–100	36	J

3.3   Problems Encountered at Pressing

There may be problems starting and shutdown of the production. First the pressing die should reach the required temperature (80–90 °C) in order to produce quality pellets. This requires time and in the warm-up period the pellet produced is of inferior quality and it should be returned to the raw material.

The shutdown of production also requires caution and care. If it is not done properly the material may stick in the boreholes with serious consequences. The cleaning of boreholes is a difficult task and, in extreme cases, the die must be replaced. Before the shutdown additives (vegetable oils) are fed into the pressing die to prevent sticking and to make a restarting possible. Most large-scale producers, use continuous production in three shifts to avoid the problems of starting and shutdown.

In large scale production an information system is mandatory. Any malfunction or deviations from the setting values of operational parameters can be detected in time and immediate intervention can be carried out.

In general, several pressing machines work in parallel. An unexpected breakdown will result in a partial loss of the production capacity and the whole production line will not be shut down.

A preventive maintenance system is also important and the main parts will be replaced after reaching their expected service life and not after each breakdown. A properly designed monitoring system gives all information on the state of the working ability of the whole production line to allow taking preventive measures in time.

The appropriate operation of a pelletizing plant requires the accurate preparation of raw material and a controlled flow of the material to the press. The latter is especially important to ensure a trouble free operation and quality product. In order to have a continuous operation independently of previous preparation stages, it is worth using a buffer bin. The discharge of material from the bin occurs with a rotary valve into the feed throat of a screw conveyor serving as a solids metering device. It is able to deliver the prescribed amount of material to the press in the unit time.

Figure 3.3 shows a drop-through valve and its arrangement on a feed hopper. The throughput of a rotary valve is calculated as

$$V_m = V_r \cdot n \cdot \Phi \cdot 60 \ \ \text{m}^3/\text{h}$$

where

V_r—is the volumetric capacity of rotor, m^3,
n—is the rotation speed of the rotor, rpm,
Φ—is the filling of the rotor, decimal.

The rotary valve is a displacement device and the quantity of material fed is proportional to the rotation speed up to a given critical speed. Above a given rotation speed, there is not sufficient time to fill and discharge the pockets of the rotor completely and, therefore, the throughput decreases compared to the theoretical value. The filling of the rotor depends on the flowing properties of the material and rotation speed.

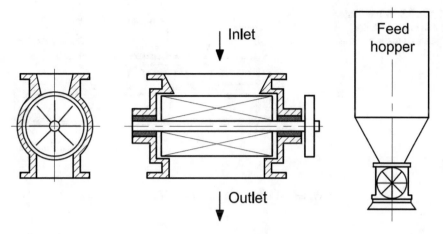

Fig. 3.3 Drop-through rotary valve and its arrangement on a feed hopper

As a guide, Fig. 3.4 shows the variation of filling of a rotary valve using different sawdust. The smaller the particle size, the lower the filling of the rotor. Due to the high variability in the flow properties, it is important to check the throughput with simple measurement. Any change in the material properties requires a change in the rotation speed of the rotor.

The assembly and working principle of a screw feeder is shown in Fig. 3.5. The material fed into the feed section is carried over the shaft. At high rotor speeds there may not be enough time to fill the feed section and less material is conveyed. That means, a filling efficiency should be taken into account. By limiting the flow area, the flow control section determines the amount of transported material per revolution. The conveying section has an increased pitch to prevent material compaction and increased friction.

The throughput of screw conveyor can be calculated by the following equation

Fig. 3.4 Filling of a rotary valve as a function of rotation speed 1—sawdust 300 kg/m^3, 2—sawdust 150 kg/m^3

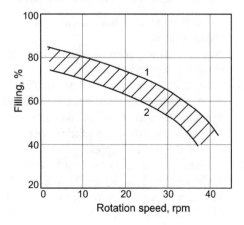

Fig. 3.5 Screw feeder
assembly with the
characteristic sections

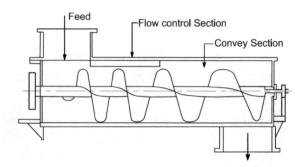

$$V_m = \frac{\pi}{4}\left(D^2 - d^2\right) \cdot h \cdot n \cdot \Phi \cdot 60 \ \text{m}^3/\text{h}$$

where

D—is the screw blade diameter,
d—is the shaft diameter,
h—is the pitch,
Φ—is the filling efficiency.

The prepared material is highly cohesive with unfavorable flow properties and the outflow from a bin can generally be ensured by the use of flow aid devices (Sitkei 1986). Therefore, a flow aid device is incorporated into the silo or hopper design in order to provide a controlled amount of discharge from the silo.

The use of flow aid devices means additional costs for their installation but they may also be used to reduce the overall cost of storage equipment. The free outflow of the material requires a conical bottom section which, in the case of small cone angle, considerably increases the height of the silo and also its cost. The use of a flow aid device enables an almost flat bottomed silo with much lower height.

The most common flow problem is the "arching" (bridging) of the material above the outlet orifice. In some cases the formation of a channel may occur through the solid mass (ratholing). The promotion of flow can be initiated by

– fluidization or aeration which involves the mixing of air to the solid phase to facilitate flow,
– air assisted promotion of flow in which high-pressure air blasts are used to break down arching and dead areas,
– a vibration device activated pneumatically or electrically and fitted externally to the conical hopper. An-other solution is the bridge breaker device which induces vibrations into the bulk material in the vicinity of bridge formation,
– hopper liners with a low friction coefficient which facilitate and promote a mass flow in the conical hopper,
– a mechanical device promoting flow which is installed into the silo designed specifically for the given device. There are different types of mechanical devices such as rotary ploughs, screw activators, paddle feeders and screw dischargers with circular motion.

In some cases, especially in large scale production, one or several pneumatic transport sections are included in the production line (Fig. 3.9). The pneumatic transport of bulk materials has been used for a long time and the basic regularities were described in the early 1950s (Pápai 1954; Pattantyus 1954). The total resistance of the system (pressure drop) has been successfully separated into two main parts: the air resistance and the resistance of material flow. Based on the theory of Pápai, who used wheat in his experiments, the corresponding resistance coefficients were later determined for sawdust and chips (Boronkai 1993). In the early studies it was theoretically derived and experimentally verified that the slip of particles, defined as

$$S = \frac{v - c}{v} = \frac{w}{v} \quad and \quad c = (1 - S) \cdot v$$

is constant and its value is around $S \approx 0.4$ independent of pipe diameter.

where

v—is the air velocity,
c—is the particle velocity,
w—is the relative velocity between air and particle.

The specific weight of the particles and the mass load somewhat modify the slip of particles. Due the different velocity of air and particles, we distinguish mass load ratio μ and mass transport concentration μ_t.

$$\mu = \frac{m_s}{m_a} \quad and \quad \mu_t = \frac{\mu}{1 - S}$$

where

m_s—is the mass load in the unit time,
m_a—is the air mass flow in the unit time.

For quick estimate of the expected pressure drop of the system with dilute mass transport is worth to use, also today, the old Gasterstädt's equation

$$\Delta p = \Delta p_0 (1 + k \cdot \mu)$$

where

Δp_0—is the pressure drop for air flow alone,
K—is a constant and its value for horizontal transportation is 0.3.

For a more accurate dimensioning of a system all resistance components must be taken into account.

In order to have a trouble-free operation, the minimum air velocity should properly be selected. Based on experimental results, Table 3.3 gives approximate values for ensuring clogging-free operation.

Table 3.3 Minimum air velocity

Type of chips	v_{min}, m/s
Dry wood dust	10–12
Sawdust, dry	12–14
Sawdust, moist.	15–17
Planer chip, dry	15
Planer chip, moist.	20
Match-stick	16

At the final stage of pneumatic conveying particles must be separated from the gas flow. A suitable system is determined by the particle size distribution, degree of separation required and the potential of transported material to pollute the environment. Cyclone separators are most commonly used but with a high content of fine particles fabric filters with mechanical shaking are also used to prevent environmental pollution.

The principle of separation of a particle in a centrifugal field is the following. If the centrifugal force on a particle is higher than the drag force, then the particle moves toward the wall and, at the same time, will move down due to gravity. On the contrary, if the drag force is higher than the centrifugal force, then the particle will move with the air toward the outlet and get into the environment. The ability of a system to separate a particle of given size is characterized by the fractional separation efficiency

$$\mu_{fr} = \frac{\Delta m_{in} - \Delta m_{out}}{\Delta m_{in}}$$

where

Δm_{in}—is the amount of particles in a given size entering the filter,
Δm_{out}—is the amount of particle in a given size leaving the filter.

The fractional separation efficiency is characterized by a curve as a function of particle size, Fig. 3.6. The particle size distribution generally follows a logarithmic normal distribution given by the following equation

$$y = \frac{1}{\sqrt{2\sigma} \cdot \bar{X} \cdot \sigma \cdot e^{\sigma^2/2}} \cdot e^{-(\ln x/\bar{X})^2/2\sigma^2} \tag{3.1}$$

where

$\bar{X}$—is the mode of distribution,
σ—is the standard deviation.

The separation efficiency shown in Fig. 3.6 can be described as follows

$$\eta_{fr} = 100\left(1 - e^{-(x/X_k)^n}\right)\% \tag{3.2}$$

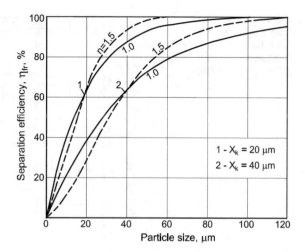

Fig. 3.6 Different separation efficiency curves

where X_k means the characteristic particle size at which the value of function is $(1 - 1/e) = 63.21\%$.

The amount of dust leaving the filter is calculated as

$$\Delta Q = (1 - \mu_{fr}) \cdot y$$

and using Eqs. (3.1) and (3.2), yields

$$\Delta Q = \frac{100}{\sqrt{2\pi} \cdot \bar{X} \cdot \sigma \cdot e^{\sigma^2/2}} \int_0^{\infty} \exp\left\{-\left((\ln x/\bar{X})^2/2\sigma^2 + (x/X_k)^n\right)\right\}\% \qquad (3.3)$$

The use of Eq. (3.3) highly facilitates the quick selection of a filter for a given particle size distribution. It turned out that the characteristic ratio $X_k/\bar{X}$ uniquely determines the amount of dust not separated in the filter. The results of systematic calculations are depicted in Fig. 3.7 (Sitkei 1994). For example, if we have a dust distribution with $\bar{X} = 250\ \mu m$ and $\sigma = 1.1$, the allowed dust pollution is 0.25% then $X_k/\bar{X} = 0.1$ and the filter should have a value of $X_k = 25\ \mu m$. A cyclone separator can fulfill this requirement. A fine dust due to sanding may have $\bar{X} = 35\ \mu m$ and $\sigma = 0.85$ values, with the allowed dust pollution of 0.25%, the filter should have $X_k = 3.5\ \mu m$ (fabric filter).

Figure 3.8 shows the strong influence of the characteristic ratio $X_k/\bar{X}$ on the separation efficiency. This method is also suitable to estimate the particle size distribution of dust not separated in the filter which pollutes the environment (Sitkei 1994).

Fig. 3.7 Amount of dust leaving the filter after separation as a function of standard deviation for different $X_k/\bar{X}$ ratios

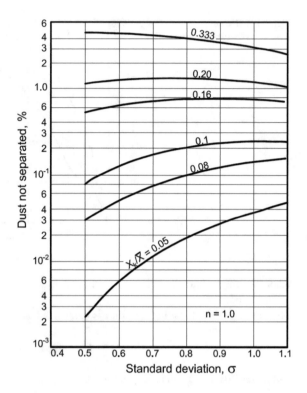

3.4   Post Treatments and Packing

The pellet coming out of the press is warm with temperatures around 80–90 °C. In the warm state, the pellet is sensitive to mechanical actions and, therefore, it should be cooled down to environmental temperature. The warm pellet should be handled with care with the least movement possible to avoid dust formation and disintegration. The cooling of pellets occurs in a counterflow cooler to avoid a high temperature difference (heats shock) between the pellet and cooling air. The warm pellet enters the cooler at the top and moves down by gravitation, and gradually cools down to the temperature of the environment. The cooling air moves upward and its temperature increases depending on the air velocity. Calculating with an initial pellet temperature of 90 °C, and a temperature of cooling air of 20 °C, the required amount of cooling air is between 8000 and 10,000 m^3 per metric ton of pellets.

Another possibility to use a conveyor belt cooler with multiple crossflow which may provide a more gentle cooling process with less mechanical damage. In small scale production natural cooling may also be used by spreading the pellets in a layer.

The dust fraction resulting in the cooling process must be removed which can be simply done with screening. The finished pellets can be stored in bulk or in bags depending on the market requirements. The smaller bags are usually 15–25 kg and the big bag contains 1.0–1.2 ton of pellets. The small bags are stored on pallets which

Fig. 3.8 Dust leaving the filter after separation as a function of the characteristic ratio $X_k / \bar{X}$

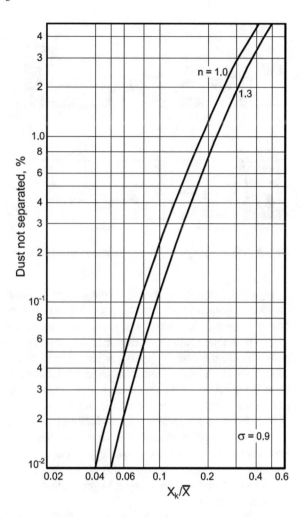

are suitable for shipping. Tankers transport bulk pellets. The loading and unloading is done mostly with a conveyor.

3.5 Technology of Pellet Production

A production technology consists of a sequence of operation processes and the plant layout follows the sequence of processes. The previous sections briefly discussed the main processes of pellet making. In order to ensure a continuous material flow, the individual processes are linked together with transport equipment (feeders). The main problem may be the synchronization of processes to a common speed which would produce the highest possible production rate and the maximum economy.

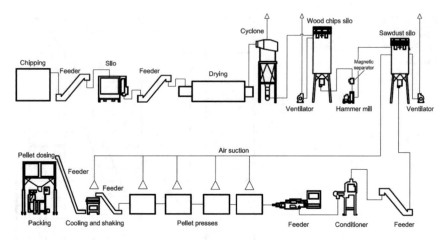

Fig. 3.9 The technology of pellet production (Burján 2009)

Figure 3.9 shows the general layout of a pellet production plant. Several storage bins are included in the technology serving as buffer capacities to keep the material flow in balance independently of shorter deviations from the nominal value.

The production capacity is determined by the average specific material flow (metric ton/h) and the working hours per year. The utilization of working time also considerably influences the production capacity. Efficient working time can be achieved by well organized production management, material supply, process control, safety of operation, preventive maintenance and product quality control.

In order to run economic pellet production, a reliable material supply with constant properties is very important. As we saw in Sect. 2.11, the selection of die ring parameters, especially the channel length, must correspond to the properties of the raw material. If the properties of the raw material change, then the die ring must also be changed which causes a break in production and an additional cost for buying different die rings.

The lifetime of a die ring is approximately 2000–2500 operating hours after which the worn parts must be replaced.

3.6 Characteristics of Wood Pellets

The size of wood pellets depends on the production technology and the requirements of the consumers. The external surface of wood pellets is smooth, free from cracks, with a maximum moisture content of 12% according to the standard. In reality, this value is typically around 7–8%. The average density of wood pellets is 1100 kg/m^3, and its bulk density is generally 650–700 kg/m^3. Its average calorific value is 17.5 MJ/kg. The pellets are mainly produced from wood and agricultural by-products. However, there are also more specific types, such as eco-pellets. Some of

Table 3.4 General characteristics of wood pellet (Kocsis 2015)

Pellet size	Diameter 6–8 mm Length 10–30 mm
Pellet heating value	16.5–18 MJ/kg (4.7–5.0 kWh/kg) ca. 3 MWh/loose m^3
Pellet moisture content	7–8%, maximum 12%
Ash content	Below 1%
Materials	Sawdust and chips, piece by-product
Pellet loose density	650–700 kg/m^3
Pellet density	1050–1200 kg/m^3
Space requirement	ca. 1.5 m^3/t
Comparison with other materials	1 m^3 oil fuel ca. 2.1 t pellet 1 t oil fuel ca 2.5 t pellet 1 m^3 wood chips ca. 0.28 m^3 pellet (0.18 t pellet) 1 m^3 natural gas ca. 2 kg pellet

the raw material for eco-pellets is waste paper, compost or, sewage sludge. Table 3.4 summarizes the general physical characteristics of wood pellets.

3.7 Energy Balance of Pellet Production

The production of pellets always requires energy. The individual processes of production require energy, electricity and heat. On the other side, the pellet produced has a heat value which will be released in a furnace. An important characteristic of process efficiency is the ratio Δf the invested and returned energy in percent. Its reciprocal value, energy returned on energy invested (EROEI), is also often used. The pellet production includes a sequence of various processes such as chipping, drying, conditioning, post-chipping, pressing, screening and material transport which all consume energy. The energy of drying depends on the moisture content to be removed and it may decisively affect process efficiency. Figure 3.10 shows the results of model calculations as a function of percentage of moisture removed by drying. In order to achieve a more realistic picture, a thermal efficiency of 80% is taken into account for energy production.

In pellet production, the energy consumption amounts to 7–23% of the heating value of a pellet depending on the drying requirement. At the same time, the EROEI-number varies between 14 and 4. The relative energy consumption can further be decreased if no chipping and drying are needed. In this case the relative energy consumption is around 4.5% and the corresponding EROEI-number is 22. It should be noted that the above calculations do not include energy consumption due to raw material transportation, gathering and handling in the forest (Kocsis et al. 2012; Kocsis 2015). This energy consumption may also be significant, especially for long

Fig. 3.10 Relative energy
consumption of pellet
production as a function of
removed moisture at drying.
A thermal efficiency of 80%
is taken into account

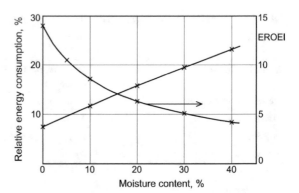

distance transportation. The source of raw material, the need for preparation, the
distance of transportation play important roles in the planning of a pellet production
system and may significantly influence the energy balance of pellet production.

The raw material from a forest generally has high moisture content with a high
drying energy requirement. If field drying is possible, the energy requirement can
be decreased. Although the energy for chipping with a knife cutter head is slightly
more, at lower moisture content, the saving from decreasing drying is overwhelming.
Furthermore, a post-chipping with a hammermill requires less energy with decreasing
moisture content. That means that, with dry raw material, it is more efficient to use
a rough knife chipping and a subsequent hammermilling. (For energy requirements
of individual processes, see Sects. 3.11 and 3.12).

3.8   Quality Requirements, Standardization

As a result of European standardization, a number of national directives have been
adopted that regulate the properties of pellets in different ways and within different
tolerances. The first pellet standards appeared at the national level at the beginning
of the 2000s. The three most important national standards are Swedish (SS 187120),
German (DIN 51731, DIN plus) and Austrian (ÖNORM M7135). Due to the unified
market and the expanding commercial activities, the application of different standards
has caused many problems. In the single market of the European Union, it is important
to have harmonized standards and at least one EU recommendation or directive are
of paramount importance. In 2003, the EU developed a technology standard adopted
on a consensual basis that includes common definitions and terminology, taking
into account national standards (CEN/TS14588 2003). In 2005, another one was
published which contained a classification for the unified characterization of fuels
(CEN/TS 14961 2005).

These standards have been further developed by the two single European standard series EN 14961 2010, which includes specifications and classifications for
solid biofuels, and EN 15234 2011, which sets quality standards for solid biofuels.

Both standards appeared as a multi-part standard and both standards included provisions for non-wood pellets. With the emergence of new European standards, national standards have been withdrawn, and replaced by localized and harmonized European standards. The main shortcoming of European standards is that they are limited to non-industrial pellets. With the release of the international ISO 17225 2014 series, which replaced EN 14961 2010, this gap has been resolved in terms of specifications and classification categories.

The standards EN 14961 and EN 15234 (mentioned above) included requirements for the moisture content and mechanical strength of the pellet, as well as testing and sampling methods. Separate chapters include general concepts, definitions, method and permissible values for the determination of moisture content, determination of ash content, sampling procedure taken during testing, sample preparation and method of calculating the calorific value. Special requirements for industrial and non-industrial pellets (wood and agri) are discussed separately.

In addition to the parameters required by the standards, there is much to say about the quality of pellets and the conditions of storage. The surface of the pellet should be glossy-, and free of long cracks, which is generally ensured by the high pressure and temperature. The healthy pellet is bright in color. The color depends to a large extent on the composition of the species of wood used and the proportion of the bark, heart and sapwood. Inadequate storage can lead to biodegradation, resulting in a change of color and odor of the pellet. The smell of healthy pellets is intense, but it will soon disappear. Inadequate storage and delivery of pellets will lead to a deterioration in their quality, so that pellets which met the standard after production may not meet the standard when they are purchased.

3.9 Energetic Utilization of Pellets

The biggest advantage of burning pellets is that they are CO_2 neutral. Of course, this is only true if we do not take into account the energy used by the plants in their lifetime and the associated emissions. This means that the CO_2 released during the combustion process is captured by the plant during its lifetime. However, when burning of fossil fuels, we add CO_2 to the atmosphere, which was previously located in the earth's crust. Thus, we influence the earth's atmospheric composition and thus the planet's climate. However, this picture is shaded by the fact that pellets are also produced and transported using additional energy, and their production burdens the environment. At the same time, pellet burning still produces about 90% less emissions than a similar gas burning (Fenyvesi et al. 2008). Pellet boilers and fireplaces are similar in design and function to other biomass combustion equipment.

Agripellets are an exception to this. Their ash has a low melting point and they produce significantly more ash compared to wood pellets. They cause increased corrosion due to their different chemical composition. All these require mechanical solutions different from boilers that burn wood pellets. High comfort and automation are one of the main reasons for the pellet's competitiveness. For newly built homes, it is advisable to allocate a separate room for storage of pellets, where bulk transport and storage are preferred. It should be taken into account that the transport vehicle may directly approach the building, so it is advisable to place the room next to an external wall. The room must be properly designed if there is pneumatic filling. A "protecting carpet" must be placed in front of the filler neck to prevent damage to the pellets and the wall during filling. The heating capacity of the built-in combustion plant and the environmental conditions of area are the basis for designing the dimension of the room. It is not obligatory to store a quantity for full year. The quantity stored depends largely on the available space, the customer's needs and transport capacity. With semi-automatic or less convenient solutions, the user empties the bag into a buffer container.

Thermal energy is released by the combustion of biomass. The combustion is a complex phenomenon involving simultaneous heat and mass transfer with chemical chain reactions and gas flow. For the sake of simplicity, the global reaction for the combustion of a biomass fuel may be given in the following form

$$C + O_2 = CO_2 + energy$$

$$2H_2 + O_2 = 2H_2O + energy$$

$$S + O_2 = SO_2 + energy$$

The following equation holds for mass balance and transformation

$$C + H_2 + O_2 + S + N_2 + u + a = 1\,kg$$

where the constituents carbon (C), hydrogen (H_2), oxygen (O_2), sulphur (S), nitrogen (N_2), moisture (u) and ash (a) must be substituted in kg.

Due to their carbohydrate structure, woody biomass contains much oxygen which considerably lowers the oxygen demand for burning. The average composition of wood fuels is the following

- carbon (C)   50% (fixed carbon, 13–16%)
- oxygen (O_2)   40–43%
- hydrogen (H_2)   5.8–6.1%
- nitrogen (N_2)   0.6%
- ash (a)   1–2.5%
- volatile matter   80–90%

The composition of fuels dictates the theoretical quantity of air needed for complete combustion. The stoichiometric air-fuel ratio for wood fuels fluctuates in a narrow range between 4 and 5 Nm^3/kg dry matter, due to their high oxygen content. Depending on the actual air flow to the burning fuel, the combustion may take place under fuel lean or fuel rich conditions which are characterized by the equivalence ratio λ (or excess air ratio)

$$V_a = \lambda \cdot V_{St}$$

where V_a, V_{St} are the actual and stoichiometric air volume in Nm^3/kg dry matter.

Wood fuels have a high volatile matter content which produces a two-phase combustion process. Combustion occurs both in the gas phase with the burning of volatile materials released through the pyrolysis of the wood upon heating and also in the solid phase as char oxidation. Particle size determines the influence of heat transfer on the rate of heating small particles will be heated rapidly while thicker particles heat more slowly (Ryu et al. 2006). The burning of volatiles is quite rapid which is followed by a slow oxidation of char.

The power capacity of a boiler is determined by the rate of pyrolysis and char oxidation which, in turn, depend on the rate of heat transfer and the kinetic rate of reactions. Conductive, convective and radiative heat transfer occurs between solid phases, and between solids and gas within the bed, and further between bed, walls and flames above the bed, drying, pyrolysis and char gasification sequences in the solid fuel and the reaction of volatile gases with the air. If the reaction rate of volatile gases attains a critical value, ignition takes place and the ignition front propagates into the bed. The ignition front moves downwards to dry and heat up the particles below which ensures a continuous burning process (Saastamoinen and Taipale 2000).

The progress of combustion is characterized by the ignition front speed and the burning rate which is the mass loss rate of the bed per unit area and unit time. The mass loss rate is proportional to the heat release rate. The air flow rate is also an important process parameter which determines the oxygen supply and convective heat transfer.

At small air flow rates the speed of ignition front is controlled by the oxygen supply. At higher air flow rates the ignition front speed is limited by the reaction rate of the fuel. A further increase of air flow rate may extinguish the flames due to the cooling effect of the fresh air.

Due to the very complex nature of heat and mass transfer, and reaction processes, we can rely only on experimental measurement. Figure 3.11 illustrates the layout of a fixed bed furnace. The fuel wood with some particle size is filled onto the grate to a given height h.

The fuel bed is ignited at the surface and the ignition front starts to propagate downwards into the bed. The air is in counter flow with a given specific mass velocity. The mass of the fuel on the grate is given as

$$m_f = \rho \cdot A \cdot h = \rho V$$

where

ρ—is the bulk density of the fuel,
A—is the cross-section of the fuel bed,
h—is the height of the fuel mass,
V—is the volume of the fuel mass.

Fig. 3.11 Schematic of a
fixed bed furnace

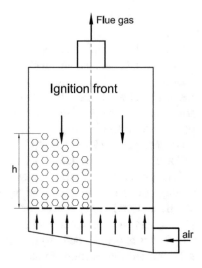

The ignition front propagates with a velocity v_i and the burning volume V_x and mass
m_x at a given elapsed time t is calculated in the following simple manner

$$V_x = A \cdot v_i \cdot t$$
$$m_x = \rho \cdot V_x = \rho \cdot A \cdot v_i \cdot t$$

And the ignition rate is given by

$$\frac{m_x}{A \cdot t} = \rho \cdot v_i \quad \text{kg/m}^2\text{h} \tag{3.4}$$

The ignition rate can be measured in model furnaces and, using Eq. (3.4), the velocity
of the ignition front can be calculated. It is an important process parameter which
can be used for a rough estimate of the expected power of a given furnace.

The true power performance of a furnace is determined by the *burning rate* of the
fuel which is not identical with the ignition rate. The burning is delayed compared
to the ignition, mainly due to the slow burning of the char. The burning rate is much
more connected to the mass loss of the fuel which can also be measured by weighing
of the model furnace during the combustion process. When most fuel, finally the
char, is consumed, the time of burning t_b can be determined. The average burning
rate B is given by the equation

$$B = \frac{m_f}{A \cdot t_b} \quad \text{kg/m}^2\text{h} \tag{3.5}$$

Pellets have heating values of 17–18 MJ/kg. Knowing the heating value of the fuel
H and the efficiency of the furnace η_f, then the expected heating capacity P of the
furnace is

$$P = H \cdot B \cdot A \cdot \eta_f \quad \text{MJ/h} \tag{3.6}$$

and the specific fuel consumption

$$M = \frac{m_f}{t_b} = B \cdot A \quad \text{kg/h}$$

Keeping in mind Eq. (3.4), the burning rate Eq. (3.5) can be rewritten as

$$B = \frac{m_f}{A_i \cdot t_b} = \rho \cdot v_b$$

where v_b means the apparent burning velocity and it correlates with the burning time in the following simple manner

$$t_b = \frac{h}{v_b} = \frac{h \cdot \rho}{B} \tag{3.7}$$

where h is the height of the fuel bed.

The presence of oxygen is an important condition for fuel combustion. If we take a constant stoichiometric ratio

$$\frac{V_a}{m_f} = 4.5 \quad \text{Nm}^3/\text{kg}$$

the corresponding air velocity through the grate is calculated as

$$v_a = \frac{4.5 \cdot m_f \cdot \lambda}{A \cdot t_b} = \frac{4.5 \cdot \lambda \cdot m_f \cdot B}{A \cdot h \cdot \rho} \quad \text{m/s} \tag{3.8}$$

or the air flow rate

$$Q_a = \rho_a \cdot v_a = \frac{4.5 \cdot \lambda \cdot \rho_a \cdot m_f}{A \cdot t_b} = \frac{4.5 \cdot \lambda \cdot \rho_a \cdot m_f \cdot B}{A \cdot h \cdot \rho} \quad \text{kg/m}^2\text{h} \tag{3.9}$$

Example Pine wood is cut into cubes 20–30 mm, with a bulk density of 290 kg/m^3, the grate surface is 0.049 m^2 (25 cm dia.), the fuel bed is 38 cm high, $m_f = 5.41$ kg, burning rate is $B = 210$ kg/m^{2}h at $\lambda = 1.0$. From Eq. (3.8) the air velocity is $v_a = 946$ m/h $= 0.26$ m/s, the air flow rate is $Q_a = 1136$ kg/m^{2}h. The net heat release with a heating value of 18 MJ/kg is 3780 MJ/m^{2}h $= 1.05$ MW/m^2. Using wood pellets, the burning rate will not be much higher and the expected values are around 240 kg/m^{2}h. The attainable power capacity is 1.2–1.3 MW/m^2.

It is important to note that the ignition, Eq. (3.4), and the burning rate as a function of bulk density of the fuel are nearly constant or slightly decrease with increasing bulk density, that is

$$\rho \cdot v_i \cong const. \quad \rho \cdot v_b \cong const.$$

As a rule of thumb, the burning rate is 65% compared to the ignition rate

$$\rho \cdot v_b \approx 0.65 \cdot \rho \cdot v_i$$

The ignition speed depends on the excess air ratio λ. In fuel rich conditions the ignition speed is limited by the availability of oxygen. In this case the carbon monoxide content in the gas phase may have high values (20–25%) due to the insufficient oxygen supply and it will burn later when the slow oxidation of char consumes less oxygen. The overall burning time naturally increases.

The moisture content of the wood particles considerably influences their ignition speed. Energy is needed to heat and evaporate the moisture and this decreases the temperature in the ignition front which adversely affects the ignition speed. Furthermore, the heating value of the fuel decreases as moisture content increases. There is an upper limit of moisture content at which self-supporting combustion is still possible. This limit is around 65% moisture content (w.b.). When burning wet fuels, carbon monoxide and other products of incomplete burning may appear in greater quantities. The approximate relative decrease of ignition speed is given in Fig. 3.12 which can be described with an exponential function in the following form

$$v_i/v_{i0} = \exp(-u/u_0)$$

where u is the moisture content, wet basis, and u_0 is the characteristic moisture content at which the function value is $1/e = 0.3679$. For this particular case $u_0 = 25\%$.

Fig. 3.12 Relative decrease of ignition speed as a function of moisture content

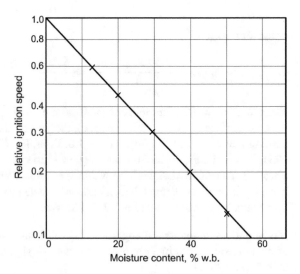

In practical cases there are several possibilities to enhance the ignition and burning speed. For instance, using a moving grate, the fuel particles also make some movements relative to each other or even some vibration in a given layer, which can increase the velocity of ignition and burning.

An important issue for all combustion equipment is the emission of pollutants. The primary pollutants formed during the combustion process are the following

- particulate matter,
- products of incomplete combustion,
- oxides of nitrogen,
- heavy metals if treated and painted woods are present.

Particulate matters include ash, condensed tars and oils, and soot. Most of these particles are less than 1 μm and respirable causing breathing hazard.

Products of incomplete combustion are CO and HC, volatile organic compounds and aromatic hydrocarbons (Jenkins et al. 1998). These pollutants are mostly generated by fuel reach conditions and excess moisture content which can be controlled. The oxides of nitrogen, mainly NO and NO_2, are formed from nitrogen in the fuel. NO formation can considerably be controlled by stoichiometry, similarly to spark ignition engines. The least NO is formed at the stoichiometric air-fuel ratio and with fuel lean conditions the NO formation considerably increases. In combustion under fuel rich conditions, the oxidation of CO is fast and formation of NO compete with each other for oxygen and, therefore, the production of NO is limited.

Combustion of various fuel types produces different amount of NO. In general, woody fuels generate less NO than straw materials do. Pellets burn at a lower furnace temperature compared to HC fuels which also reduces NO formation

3.10 Competitiveness and Market Conditions

The market demand for pellets depends on many factors but its price related to the energy content (\$/MJ) is fundamental. This price should be compared to those of natural gas and heating oil provided that they are also available at a given locality. The picture is further complicated by the fact that a government may control the price of natural gas and heating oil, for social reasons but much less that of pellets.

Comparing the heating values, 1 kg oil (46 MJ) is equivalent to 1.33 m^3 of natural gas and 2.55 kg of wood pellets. For example in Hungary, the price of 100 MJ energy from the different energy sources are \$2.37 for oil, \$1.14 for gas, \$1.42 for pellets and \$2.83 for electricity. Natural gas is the cheapest energy source and here are no costs for delivery and storage. Of course, not all areas have natural gas and in this case pellets may be competitive with heating oil. Furthermore, the density of population influences the economical use of piped natural gas supply and this may also promote the use of pellet firing.

A further important point that only the cost of pellet making is variable and it may drop below the cost of the natural gas. A further important question is the availability of raw material for pellet making. Bigger countries with good forest cover have the potential resource to run large scale pellet production with a reliable raw material supply. Smaller countries with less forest are limited to medium size and small scale pellet production based rather on industry wastes and by-products. A big question is the use of agricultural by-products, mainly straws and maize stalks including energy plants. After harvesting many plant residues must be removed from the field to ensure the appropriate planting conditions for the next crop. The use of these by-products is still not economical. Some 50 years ago there were trials to burn big round bales in special furnaces.

Agricultural by-products have some disadvantages because of their ash properties (Jenkins et al. 1998). Therefore, the maximum gas temperature in a furnace is limited to avoid ash melting and deposition.

Pellet production is continuously increasing in the world (Fig. 3.13). The main producers are North America, Western Europe followed by South America and Eastern Europe. Recently China has also increasing production. Because China is short of forests, most of industry wastes (sawdust) and by-products are used. China has agricultural by-products and residues in abundance and, therefore, the agripellet production more promising. A characteristic difference between Western Europe and China is that in Western Europe pellets are mostly used by households, while in China by the industry and power plants.

The world has tremendous sources of wastes, by-products and residues. A major task is to use these materials in an economical way for the benefit of mankind. New processing methods and technologies may promote the further increase of pellet

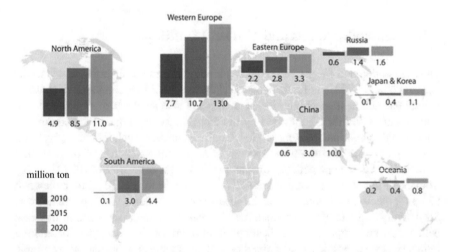

Fig. 3.13 The pellet production of the world in 2010 and 2015, and the forecast in 2020 (http://www.siteselection.com)

production world-wide. Large scale production will probably be limited to certain bigger countries with large forested areas. Nevertheless, in many countries the small scale producers also have a chance to use the local wastes and by-products. The economical use of agricultural by-products and residues requires further developments both in the processing and burning technologies.

3.11 Some Practical Observations on Energy Requirements

Laboratory measurements are very useful to clear and establish general relationships describing system behavior as a function of influencing variables. At the same time, machines and equipment have auxiliary mechanisms which also consume energy. That means that in practice energy consumption does not equal those amounts determined in the laboratory. Therefore, practical observations usefully complete our knowledge obtained in the laboratory.

In Fig. 3.14 we summarized the practical observations concerning the specific compaction energy as a function of pellet diameter. It can be seen that the laboratory measurement values slightly underestimate the practical values. Furthermore, in the practical observations we can not always select coherent values for the same material for different pellet diameters. As a consequence, a scattering zone should always be reckoned with. Nevertheless, the practical observations follow well the more exact laboratory measurements.

A common practice in the preparation of chips for pellet making is to pre-grind raw material into larger sizes of 10–20 mm and followed by fine-grinding using hammer mills. The general regularities and operational characteristics of hammer mills for agricultural products (maize, barley) have already been investigated in the early 1960s and their performance characteristics are well-established (Sitkei 1986). The governing laws and processing methods elaborated here are also suitable to determine the performance of hammer mills for grinding wood materials. In the following we demonstrate this method for pine wood as an example. In a comminution process,

Fig. 3.14 The specific pressing energy as a function of pellet diameter

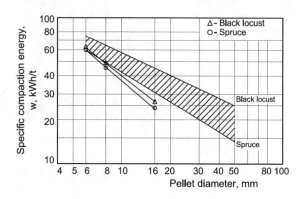

the initial surface area of a particle will be increased (Δf, cm²/g) and the increase in specific surface area consumes energy. The specific energy of comminution (v, kWh/cm²) depends on several factors such as timber species and their mechanical properties, moisture content and slightly also on the geometric size distribution.

The net power consumption of comminution, without the idling power, is given by the simple equation as

$$P_e = v \cdot \Delta f \cdot 10^6, \quad \text{kWh/t}$$

There are several practical variables such as the

– geometric mean of particle size distribution,
– screen size used in the hammer mill (generally 1–10 mm)

which have strong interrelations with the increase in specific surface area Δf.
Figure 3.15 shows the relationship between the required surface increase and the geometric mean d_g supposing an initial mean particle diameter of 5 and 10 mm. This Figure also shows the specific energy of comminution calculated from the net power requirement.
The specific energy of comminution is roughly twice as high as those measured for maize and barley.
The curves in Fig. 3.15 can be described with the following simple empirical equations

$$\Delta f = 105 \cdot d_g^{-1.15}, \quad \text{cm}^2/\text{g} \tag{3.10}$$

Fig. 3.15 Increase in specific surface area and specific energy of comminution as a function of mean geometric diameter. Moisture content 12%. 1—d_i= 5 mm, 2—d_i= 10 mm

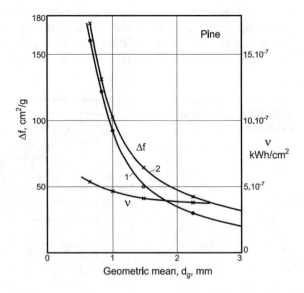

$$v = 4.8 \times 10^{-7} \cdot d_g^{-0.25}, \quad \text{kWh/cm}^2$$

and further

$$P_e = v \cdot \Delta f = 50 \cdot d_g^{-1.4}, \quad \text{kWh/t} \tag{3.11}$$

where the geometric mean d_g must be substituted in mm.

The screen size used in a hammer mill strongly influences the increase in surface area Δf and also the power consumption. Figure 3.16 shows the influence of screen size on the surface increase during comminution and the corresponding power consumption.

The following equation holds

$$\Delta f = 164 \cdot d_{sc}^{-0.52}, \quad \text{cm}^{2}/\text{g}$$

which, combining with Eq. (3.10), yields

$$d_g = 0.68 \cdot d_{SC}^{0.45}, \quad \text{mm} \tag{3.12}$$

where the screen size d_{SC} must be substituted in mm.

The moisture content of raw material has a significant effect on the specific energy of comminution and the specific power consumption. Quite similarly to crop products, the power consumption increases with the moisture content in a parabolic fashion and Eq. (3.11) can be supplemented in the following form

$$P_e = 50 \cdot d_g^{-1.4} \left[1 + 0.33(U - 12)^{0.5} \right], \quad \text{kWh/t} \tag{3.13}$$

where the moisture content U (w.b.) is substituted in per cent.

It should be stressed that the numerical values given in the above equations refer to a pine wood with a density of around 500 kg/m³. Hardwoods require more energy for comminution compared to soft woods. Within one species, its density may have a definite effect on the required energy.

Fig. 3.16 Interrelation between increase in surface area, power consumption and screen size. Pine, moisture content 12%

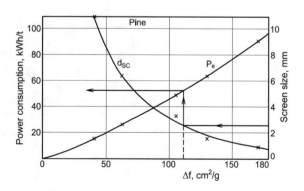

The initial particle size has also some influence on the increase in specific surface area Δf, as shown in Fig. 3.15. Its relative influence, decreases toward smaller geometric means.

In agriculture, biomass residues such as barley, oat and wheat straw represent an abundant, inexpensive source of renewable energy. They can easily be adapted in direct combustion or co-firing with coal. For grinding straws, the hammer mill is the most common technique due to its ability to grind a wide variety of raw materials. Also in straw milling, the screen opening size is the most significant factor affecting the mill performance. Higher moisture content would increase the energy consumption and, therefore, it should not exceed 12–15%. The expected energy consumption may be estimated by the following simple equation

$$P_e = 9.33 \cdot d_g^{-1.9}, \quad \text{kWh/t}$$

which shows the strong influence of geometric mean on the grinding energy.
Using screen size around 6 mm, a geometric mean of around 1 mm can be achieved with an average energy consumption of 10–12 kWh/t.

Figure 3.17 shows the energy requirement and throughput of briquette production as a function of final density. The briquette size is between 40 and 60 mm. Screw presses have somewhat more friction losses and therefore, they require more energy compared to piston presses. The energy consumption increases proportional with increasing density and the operation over 1.1 g/cm^3 density is not economic. At the same time, the throughput decreases considerably as the briquette density increases. The durability of the briquette increases with higher densities.

Briquettes are rather suitable for manual feeding with additional labor. Pellets from 6 to 8 mm in diameter are suitable for automatic feeding device with greater convenience and higher investment. Wood bark is often used for briquettes with

Fig. 3.17 Specific energy of briquetting and throughput as a function of briquette density, size between 40 and 60 mm

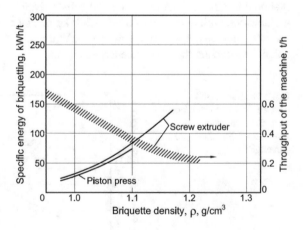

lower chipping requirements and the total production cost may be lower compared to woody pellets. A special case is the billet-size briquette production made of bark and it is used quite similarly to firewood.

3.12 Economy of Pellet Production

Today most of the raw material used in pellet production is woody biomass. This woody biomass comes directly from the forest (thinning, other by-products) or from saw mills and factories manufacturing wood products. The manufacture of wood pellets may be profitable and pellets have become a successful product traded nationally and internationally with a growing market size. Large scale pellet production, however, requires larger wood sources both from forests and wood industry.

The heating value of woody pellets must compete with other energy sources such as electricity, gas and oil. Moreover, the latter energy sources require simpler heat generation equipment providing a complete and efficient burning. Woody pellets have also an enhanced heating value in the unit volume and there are boilers specially designed for using pellets with automatic feeding devices. This equipment ensures a complete and efficient burning with low ash content and particulate emission. The use of pellets varies for heating in small-scale residential and public buildings and also co-firing in coal power plants.

The economic use of pellets is determined by the pellet production costs which depend on many influencing factors and conditions. The main cost elements are the cost of raw material and its preparation costs (chipping, grinding or hammermilling, drying, conditioning), industrial equipment (chipper, hammermill, dryer, pellet mill, cooler, screener and bagging), peripheral equipment (feeder, conveyor, loader) and operating costs such as labor costs and consumables.

The collection, storage and transportation of the raw material to the factory is an important cost factor. Depending on the wood species and the distance of transportation, the cost may vary between \$40 and \$60/t. If waste products in a factory are used (sawdust, edging) then the cost of raw material may be much less. Many of the processes require energy. If the raw material is debarked (roundwood in green condition) then the energy of chipping is 50–65 kWh/t (180–230 MJ/t). Post-chipping or hammermilling requires, on average, 15 kWh/t but hammermilling may require more energy if a smaller screen size is selected (see Figs. 3.14 and 3.15).

Green raw materials require drying which consumes much energy depending on the percentage of water removed. Each per cent of water removed requires 55 MJ/t specific energy. If 30 or 40% (w.b.) moisture should be removed, then 1650 or 2200 MJ/t energy consumption must be reckoned with. The moisture content after drying may not be uniform and drying generally occurs below the optimum moisture content (10–12%). Therefore, the material is often conditioned with saturated steam. Moreover, the steam softens the fibers increasing the durability of the pellets. Each per cent rewetting requires 10 kg water/t. The enthalpy of saturated steam is 2.68 MJ/kg and, therefore 1% rewetting theoretically requires 26.8 MJ/t energy. In practice, taking into account the efficiency of boiler, this energy is 38 MJ/t. Generally, from 3 to 5% rewetting is required with energy consumption of 114 and 190 MJ/t respectively. Pellets with diameters of 6 or 8 mm are used. The energy of pressing in this range varies between 50 and 90 kWh/t (180–320 MJ/t).

Summing up these energy requirements, the total energy consumption may vary between 2100 and 2600 MJ/t. Taking an average cost of $0.07/kWh for heat and electricity, the expected cost amounts from $40 to $50/t. If no drying and conditioning are needed, the expected energy consumption is around 750 MJ/t with a cost of $15/t.

The capital investment is the main cost factor of a pellet making factory. The common layout for pellet production technology using green raw material was already depicted in Fig. 3.9. Depending on the capacity of the particular equipment (chipper, hammermill, dryer, pellet mill, boiler, cooler, bagging system), the costs of investment may be quite different. As a guide, for a 75,000 t/year production capacity, working in three shifts with 10 people per shift, the total installed costs were around $12 millions (Pirraglia et al. 2010). That means an investment of $160/t capacity per year. Again, if no drying and conditioning are needed, lower investment costs may be reckoned with. The number of working shifts per day considerably influences the investment cost.

In order to maintain constant net present value of a factory, the expected internal rate of return (IRR) around 10 or 12% should be taken into account in the price calculation and it will be added to the net production cost. Taking the above investment cost of $160/t, a 12% IRR amounts to $19.2/t additional cost. A further revenue margin is required both for the producer and the retailers.

The operating costs include the workers wages and benefits, the different consumable costs (bags, pallets, packing materials) and the costs of parts and replacements, marketing and sales fees. The total number of workers includes the production workers, supervisor, forklift operator and maintenance technicians and it depends on the production rate. For example, a capacity of 70,000 t/years with two separate production lines requires at least 10 workers per shift. The direct labour cost varies between $10/h and $15/h and the operating cost may fluctuate, depending mainly on the consumable costs (selling in bags or in bulk), between $50/t and $80/t.

The total production costs consist of raw materials, energy, and operating costs while the selling price, contains the depreciation, taxes and the revenue for producer and retailers. Figure 3.18 shows the main costs of wood pellets using green roundwood which must be dried and conditioned. The production cost is sensitive to changes in raw material cost, labour cost, and drying and conditioning costs.

The production capacity per year influences the production cost, Fig. 3.19. For purchased green raw material, the average production cost is around $200/t at a production capacity in the range of 60,000 t/year. An increase of production capacity slightly lowers the production cost. Working in one shift, due to constant costs and relatively higher force numbers, the decreased production rate increases the production cost. Using dried raw material, needing no drying and conditioning, the production

Fig. 3.18 The main cost
elements of wood pellets
(Raw material is green
roundwood)

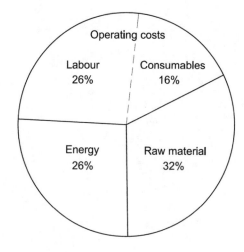

Fig. 3.19 Effect of
production capacity on the
production cost 1—green
roundwood, 2—dried
roundwood and wastes,
3—own dried and mostly
chipped raw material

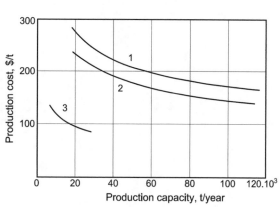

cost will be lower independently of production capacity. Using its own sawdust and
edging materials in a dried condition, the production cost may be considerably lower
but generally at a low production capacity due to the limited availability of raw
material.

It may be concluded that the pellet production is a viable activity with good per-
spectives. Depending on the raw material properties and availability, quite different
production capacities may economically transform local by-products and residues
into useful energy source which is fully competitive with other energy sources.

Concluding Remarks

In the world there are tremendous sources of wastes, by-products and residues of plant origin. There is a need to use this biomass in an economical way for the benefit of mankind. Reliable technologies have been developed for using wood by-products to make pellets and briquettes which mainly fuel boilers for heat generation. Agricultural by-products, which are also abundant in many countries require further development of appropriate technologies to use them economically.

Pellet production requires a properly prepared chipped material. The preparation of raw material mainly depends on its original size, wood species and moisture content. The availability of raw material with more or less constant properties is crucial to use a given production line economically. A raw material with high moisture content requires much energy for drying to reduce the moisture content to the desirable range of 8–11%. Natural drying of raw material may considerably reduce the necessary time and energy needed for artificial drying. Capital investment and its related costs also depends on the raw material properties and the required preparations to a considerable extent.

The production costs of pellets and briquettes are generally competitive with other energy sources. Natural gas may be an exception depending on the government price policy in particular countries. Heating oil and especially electricity have higher prices related to the energy unit ($/MJ). Therefore, the pellet firing may be economic in many countries, especially in smaller towns and rural settlements.

Pellet firing is not only economic but also generally, environmental friendly. Using woody biomass, the emission of pollutants such as particulates, incomplete combustion products and oxides of nitrogen is low, provided that the combustion equipment operates with the right stoichiometry. The ash content is also low and generally causes no problem, such as deposits, in the combustion equipment. The whole firing and heating process can be well regulated and controlled ensuring a comfortable and economic use in households and communal buildings.

© Springer Nature Switzerland AG 2019
Z. Kocsis and E. Csanády, *Theory and Practice of Wood Pellet Production*,
https://doi.org/10.1007/978-3-030-26179-5

Literature

Akdeniz RC, Haghighat S (2013) The effect of die dimensions, raw material moisture content and particle size on pelletizing characteristics of olive cake. Department of agricultural machinery, international conferences engineering, agriculture, waste management and green industry innovation. Gödöllő, Hungary, 13–19 Oct, S, pp 33–41

Anon (2009) Pellet deal may herald new era. New Zealand Forestry Bulletin (Winter 2009), New Zealand Forest Owners, p 2

Andrew C (2004) Low carbon heating with wood pellet fuel. Report, pp 4–11

Biot MA (1954) Theory of stress-strain relations in anisotropic viscoelasticity and relaxation phenomena. J Appl Phys 25(11):1385–1391. https://doi.org/10.1063/1.172157

Boltzmann L (1876) Zur Theorie der elastischen Nachwirkung. Ann Physik und Chem 624

Boronkai L (1993) Dimensioning of suction and pneumatic transport devices in the wood industry. Ph.D. dissertation, University of Sopron, Hungary (in Hungarian)

Brunauer S, Emmett PH, Teller E (1938) Adsorption of gases in multi molecular layers. J Am Chem Soc 60:309–319

Buckingham E (1914) On physically similar systems illustrations of the use of dimensional equations. Phys Rev 4:S 345–376. https://doi.org/10.1103/physrev.4.345

Burján Z (2009) Faalapú pelletgyártás alapanyagai, gyakorlati tapasztalatok (Production of wood-pellet). InnoLignum Erdészeti és Faipari Szakvásár és Rendezvénysorozat, Sopron (in Hungarian)

Csanády E (2013) Mechanics of wood machining. Springer, Berlin Heidelberg, pp 25–30. https://doi.org/10.1007/978-3-642-29955-1

Csanády E, Magoss E, Kovács Zs, Ratnasingam J (2019) Optimum design and manufacture of wood products. Springer, Berlin Heidelberg

Demirbas A (2001) Relationships between lignin contents and heating values of biomass. Energy Convers Manage 42(2):183–188

Dent RW (1977) A multilayer theory for gas sorption. Part I. Sorption of a single gas. Text Res J 47:145–152

Dmitry T, Chander S, Mathew L (2013) Effect of additives on wood pellet physical and thermal characteristics a review. Hindawi Publishing Corporation. ISRN Forestry. Volume 2013, Article ID 876939, 6 pages. https://doi.org/10.1155/2013/876939

Escort G (2009) Wood pellets—a cleaner kind of fuel. NZ J For 53(4):8–19

Fenyvesi L, Ferencz Á, Tóvári PA tűzipellet (Pellet) (2008) Cser könyvkiadó és Kereskedelmi Kft, Budapest, 85 p (in Hungarian)

Findley WN, Lai JS, Onaran K (1976) Creep and relaxation of nonlinear viscoelastic materials with an introduction to linear viscoelasticity. Dover Publication Inc., Minnesota, pp 55–74

Hailwood AJ, Horrobin S (1946) Absorption by water of polymers. Analysis in terms of a simple model. Trans Faraday Soc 42:84–92, 94–102

Heiko T, Clovis RH, Philip EH (2005) Modelling the physical processes relevant during hot pressing of wood-based composites. Part II. Rheol 64(2):125–133. https://doi.org/10.1007/s00107-005-0032-5

Hofko B (2006) Rheologische Modelle zur Beschreibung des Verformungsverhaltens von Asphalten. Betreuer/in(nen) R. Blab, K. Kappl, Institut für Straßenbau und Straßenerhaltung, 2006, Abschlussprüfung 24.11.2006

Jenkins BM, Baxter LL, Miles TR Jr, Miles TR (1998) Combustion properties of biomass. Fuel Process Technol 54:17–46

Karwandy J (2007) Pellet production from sawmill residue a Saskatchewan perspective. Saskatchewan Forest Centre. Forest Development Fund Project 2006-29, Final report 2006/07

Kocsis Z (2015) A fa szemcsés halmazok tömörítésének rheológiája és energetikája a pelletálási tartományban. (The compression rheology and energetics of wood particles at the pelleting range). Ph.D. dissertation, Hungary, Sopron, 134 p. https://doi.org/10.13147/nyme.2015.002 (in Hungarian)

Kocsis Z, Csanády E (2015) Study of the energy requirements of wood chip compaction. Drvna Industrija 66(2):163–170. https://doi.org/10.5552/drind.2015.1409. ISSN 0012-6772

Kocsis Z, Németh G, Varga M (2012) Energy demand of briquetting and pelleting of wood based by-product. The impact of urbanization, industrial and agricultural technologies on the natural environment. In: International scientific conference on sustainable development and ecological footprint. Nemzeti Tankönyvkiadó, Nyugat-magyarországi Egyetem, Sopron, pp 393–400. ISBN 978-963-19-7352-5 (in Hungarian)

Langhaar H (1951) Dimensional analysis and theory of models. Wiley, London. S, pp 25–34

Marosvölgyi B (2011) A pelletálás és brikettálás energiamérlegének vizsgálata. (Investigation of energy balance of briquette and pellet). Műszaki tudományos füzetek. Erdélyi-Múzeum Egyesület (in Hungarian)

Matthies HJ, Busse W (1966) Neuere Erkenntnisse auf dem Gebiete des Verdichtens von Halmgut mit hohem Normaldruck. Grundl Landtechn 16(3):87–92

Mohsenin N, Zaske J (1975) Stress relaxation and energy requirements in compaction of unconsolidated materials. J Agric Eng Res 11:193–205

Molnár S (1999) Faanyagismerettan (Wood material science). Mezőgazdasági Szaktudás Kiadó, Budapest, pp 260–292 (in Hungarian)

Mózes Gy, Vámos E (1968) Reológia és reometria (Rheology and rheometry). Műszaki Könyvkiadó, Budapest (in Hungarian)

Pattantyus G (1954) Pneumatic conveying. Acta techn Hung. Budapest, pp 129–177

Pápai L (1954) Pneumatikus gabonaszállítás. (Pneumatic conveying of grains). Trans Hung Acad Sci, 319–369

Pizzi A, Bariska M, Eaton NJ (1987) Theoretical water sorption energies by conformation analysis. Part 2. Amorphous cellulose and the sorption isotherm. Wood Sci Technol 21:317–327

Pirraglia A, Gonzalez R, Saloni D (2010) Techno-economical analysis of wood pellets production for U.S. manufactures. BioResources 5(4):2374–2390

Ryu C, Yang JB, Khor AE, Yates NN, Sharifi V, Swithenbank J (2006) Effect of fuel properties on biomass combustion. Part I. Fuel 85:1039–1046

Saastamoinen JJ, Taipale R (2000) Propagation of the ignition front in beds of wood particles. Combust Flame, 123214–123226

Sacht HO (1967) Über den Verdichtungsvorgang bei landwirtschaftlichen Halmgütern und die dabei auftretende Wandreibung. Grundl Landtechn 17(2):47–52

Schofield RK (1935) The pF of the water in soil. In: 3rd international congres soil science transactions, vol 2, pp 34–48

Schwanghart H (1969) Messung und Berechnung von Druckverhältnissen und Durchsatz in einer Ringkoller-Strangpresse. Aufbereitungs-Technik 12:713–722

Sitkei Gy (1981) A mezőgazdasági anyagok mechanikája. (Mechanics of agricultural materials). Akadémiai Kiadó, Budapest, pp 126–419 (in Hungarian)

Sitkei Gy (1986) Mechanics of agricultural materials. Elsevier

Sitkei Gy (1994) A faipari műveletek elmélete (Theory of wood processing). Mezőgazdasági Szaktudás Kiadó. Budapest, pp 105–140 (in Hungarian)

Sitkei Gy (1994) Non-linear rheological method describing compaction processes. Int Agrophys 8:137–142

Sitkei Gy (1997) A non-linear viscoelastic-plastic model describing compaction processes. In: Proceedings of IMACS/IFAC conference Gödöllő, pp 105–112

Skalweit H (1938) Kräfte und Beanspruchungen in Strohpressen. RKTL-Schrift, No. 88

Stamm A (1946) Passage of liquids, vapours and dissolved materials through softwoods. USDA Bull 929

Stamm A (1964) Wood and cellulose science. Ronald, New York

Timoshenko S, Woinowsky-Krieger S (1957) Theory of plates and shells. McGraw hill Company, S, pp 37–143

Tiemann HD (1906) Effect of moisture upon the strength and stiffness of wood. US Dep Agric For Serv Bull 70:144 p

Varga M (1983) Por-forgács halmazok mechanikai tulajdonságai, különös tekintettel a tartályból való kifolyásra. (Mechanical properties of bulk wood chips). Ph.D. dissertation, Sopron, Hungary, pp 74–144 (in Hungarian)

Алферов СА (1956) Исследование процесса прессования соломы. Investigation of straw pressing) Диссертация МИМЕСХ, Москва

Долгов И, Васильев К (1967) Математические методы в земледельческой механике. (Mathematical methods in agricultural mechanics), Издательство „Машиностроение", Москва

Пустыгин М (1937) Закон сжатия слоя стеблей хлеба. (Compaction law for straw). Сельхозмашина, No. 12

Особов ВИ (1962) Теоретические к экспериментальные исследования процесса брикетирования сена, (Theoretical and experimental investigation of hay briquetting). Труды ВИСХОМ, Вып 39

Index

© Springer Nature Switzerland AG 2019
Z. Kocsis and E. Csanády, *Theory and Practice of Wood Pellet Production*,
https://doi.org/10.1007/978-3-030-26179-5